Ending War

Ending War: A Dialogue across Disciplines examines how wars end from a multi-disciplinary perspective and includes enquiries into the politics of war, the laws of war, and the military and intellectual history of war.

In recent years, the changes in the character of contemporary warfare have created uncertainties across different disciplines about how to identify and conceptualise the end of war. A whole constellation of questions arises from such uncertainties: How do philosophers define ethical responsibilities *in bello* and *post bellum* if the boundary between war and peace is ever so blurred? How do strategists define their objectives if the teleology of action becomes uncertain? How do historians bracket the known endings of war and delve into the arguments that preceded them? Which answers can international law provide for the ending of wars – and which challenges remain or have recently arisen?

This volume addresses these questions and enables both an understanding of how 'the end' as a concept informs the understanding of war in international relations, in international law, and in history, as well as a reconsideration of the nature of scientific method in the field of war studies as such.

The chapters in this book were originally published as a special issue of *The Journal of Strategic Studies*.

Chiara De Franco is Associate Professor of International Relations and Co-Director of the Center for War Studies at the University of Southern Denmark. Her research focuses on conflict warning, protection of civilians, and language in international practices. She won the 2021 ISA Best Book Award with *Warning about War: Conflict, Persuasion and Foreign Policy* (2019).

Anders Engberg-Pedersen is Professor of Comparative Literature in the Department for the Study of Culture at the University of Southern Denmark. His research focuses on the relations between war, aesthetics, and the history of knowledge. He is, among other works, the author of *Empire of Chance. The Napoleonic Wars and the Disorder of Things* (2015).

Martin Mennecke is Associate Professor of International Law at the University of Southern Denmark. His research focuses on atrocity prevention, the United Nations, and transitional justice. Among his recent books is *Implementing the Responsibility to Protect: A Future Agenda* (2020).

Ending War

A Dialogue across Disciplines

Edited by
**Chiara De Franco,
Anders Engberg-Pedersen and
Martin Mennecke**

Routledge
Taylor & Francis Group

LONDON AND NEW YORK

First published 2022
by Routledge
2 Park Square, Milton Park, Abingdon, Oxon OX14 4RN

and by Routledge
605 Third Avenue, New York, NY 10158

Routledge is an imprint of the Taylor & Francis Group, an informa business

Introduction, Chapters 2–5 © 2022 Taylor & Francis
Foreword © 2022 Jamie Shea
Chapter 1 © 2019 Cian O'Driscoll. Originally published as Open Access.

British Library Cataloguing in Publication Data
A catalogue record for this book is available from the British Library

ISBN: 978-1-032-14886-1 (hbk)
ISBN: 978-1-032-14888-5 (pbk)
ISBN: 978-1-003-24159-1 (ebk)

DOI: 10.4324/9781003241591

Typeset in Myriad Pro
by Newgen Publishing UK

Publisher's Note
The publisher accepts responsibility for any inconsistencies that may have arisen during the conversion of this book from journal articles to book chapters, namely the inclusion of journal terminology.

Disclaimer
Every effort has been made to contact copyright holders for their permission to reprint material in this book. The publishers would be grateful to hear from any copyright holder who is not here acknowledged and will undertake to rectify any errors or omissions in future editions of this book.

Contents

Citation Information

The following chapters were originally published in *The Journal of Strategic Studies*, volume 42, issue 7 (2019). When citing this material, please use the original page numbering for each article, as follows:

Introduction

How do wars end? A multidisciplinary enquiry
Chiara De Franco, Anders Engberg-Pedersen and Martin Mennecke
The Journal of Strategic Studies, volume 42, issue 7 (2019), pp. 889–900

Chapter 1

Nobody wins the victory taboo in just war theory
Cian O'Driscoll
The Journal of Strategic Studies, volume 42, issue 7 (2019), pp. 901–919

Chapter 2

How do wars end? A strategic perspective
Joachim Krause
The Journal of Strategic Studies, volume 42, issue 7 (2019), pp. 920–945

Chapter 3

In pursuit of accountability during and after war
Thomas Obel Hansen
The Journal of Strategic Studies, volume 42, issue 7 (2019), pp. 946–970

Chapter 4

The Joint Chiefs of Staff, the atom bomb, the American Military Mind and the end of the Second World War
Phillips Payson O'Brien
The Journal of Strategic Studies, volume 42, issue 7 (2019), pp. 971–991

Chapter 5

Slow failure: Understanding America's quagmire in Afghanistan
Christopher D. Kolenda
The Journal of Strategic Studies, volume 42, issue 7 (2019), pp. 992–1014

For any permission-related enquiries please visit:
www.tandfonline.com/page/help/permissions

Notes on Contributors

Chiara De Franco is Associate Professor of International Relations and Co-Director of the Center for War Studies at the University of Southern Denmark. Her research focuses on conflict warning, protection of civilians, and language in international practices. She won the 2021 ISA Best Book Award with *Warning about War: Conflict, Persuasion and Foreign Policy* (2019).

Anders Engberg-Pedersen is Professor of Comparative Literature in the Department for the Study of Culture at the University of Southern Denmark. His research focuses on the relations between war, aesthetics, and the history of knowledge. He is, among other works, the author of *Empire of Chance. The Napoleonic Wars and the Disorder of Things* (2015).

Thomas Obel Hansen is Senior Lecturer in the School of Law / Transitional Justice Institute at Ulster University (UK) since January 2016. He obtained his LLM (2007) and PhD in Law (2010) from Aarhus University Law School (Denmark). His work focuses on international criminal law and transitional justice.

Christopher D. Kolenda is Author of *Zero-Sum Victory: What We're Getting Wrong About War* and is the first American to have fought the Taliban as a commander in combat and engaged them in peace talks.

Joachim Krause is Director of the Institute for Security Policy at the University of Kiel and Editor-in-Chief of the German language journal for strategic studies, *SIRIUS*. He was Professor for International Relations at the University of Kiel from 2001 to 2016.

Martin Mennecke is Associate Professor of International Law at the University of Southern Denmark. His research focuses on atrocity prevention, the United Nations, and transitional justice. Among his recent books is *Implementing the Responsibility to Protect: A Future Agenda* (2020).

Phillips Payson O'Brien is the Chair of Strategic Studies in the School of International Relations at the University of St Andrews. Before joining St

Andrews, he worked at the University of Glasgow and Cambridge University, where he received his PhD.

Cian O'Driscoll is Associate Professor of International Relations at the Coral Bell School of the Asia Pacific, ANU. He has written extensively on the ethics of war and the just war tradition. His most recent monograph, *Victory: The Triumph and Tragedy of Just War*, was published in 2019.

Jamie Shea is Honorary President of the Center for War Studies at the University of Southern Denmark, Professor of Strategy and Security at the University of Exeter, and former NATO's Deputy Assistant Secretary General for Emerging Security Challenges.

Foreword

By Jamie Shea

It has become commonplace to decry the "endless wars" of the age in which we live. Indeed if we think of the United States and NATO intervention in Afghanistan, lasting longer than World Wars One and Two combined, or the civil war in Syria now in its eleventh year, or the repeated flare up in violence in places like Somalia, the Democratic Republic of Congo, South Sudan, Ethiopia and the Sahel, we can clearly see that not only are the conflict parties unable to achieve a decisive victory on the battlefield, but diplomats are finding it just as difficult to negotiate durable settlements. Even where ceasefires and peace frameworks can be agreed, and reconciliation processes launched, ensuring effective implementation is proving to be another headache for the international community. In short, we know how wars start, but we are increasingly at a loss to determine how they can end.

This volume usefully addresses that gap, and encourages us not to despair that we are helpless when it comes to finding solutions to even the most deep rooted and intractable of conflicts. Of course, given that many of the conflicts discussed in this volume have extended over several years and involved a variety of state and non-state actors, determining at which level of diminution of violence and progress in a peace process a conflict has durably ended is no easy task. Occasionally, although rarely, we do see wars come to an end in the old fashioned way because one side achieves a decisive military victory over the other side, incapacitating his ability to carry on fighting and dictating a victor's peace. This has recently happened in Afghanistan with the return of the Taliban to power in Kabul. Yet this victory was made possible by the voluntary withdrawal of the US and NATO forces, without suffering a military defeat, as well as by their removal of support from the local Afghan security forces. Without this withdrawal of the international security forces, the military status quo in Afghanistan would undoubtedly have continued for a long time, and we must still wait to see if anti-Taliban resistance forces will re-emerge as they did during the Taliban's previous period in power (1996–2001), or the US and its allies will intervene in Afghanistan once more to strike at terrorist training camps. So peace in Afghanistan is still far from secure. Of course, wars have also ended because of peace settlements usually imposed by the victorious power on the defeated power, and often at punitive cost. Sometimes these settlements have succeeded in bringing about a durable peace (as with Appomatox in 1865 at the end of the American civil war or Toyko Bay in 1945 at the end of the US war against Japan). Yet at other times the peace agreements have caused resentment and the desire for revenge, as with the 1871 peace imposed by Prussia upon France or the Versailles Treaty of 1919 imposed upon Germany.

It is the way that wars end that determines their place in history and impact on human civilisation. Do they secure a more just and lasting peace? Do they lead to the creation of new international institutions that help to prevent future wars? Do they remove the aggressive nationalisms, expansionist ideologies or militarist cultures that started the wars in the first place? Does humanity move on as we once hoped with the slogan: "a war to end all wars", or simply relapse sooner or later back into the same vicious cycle of violence? Wars can be judged to be just or worthwhile not only in terms of the motivations and objectives of the belligerents in taking up arms but also in terms of the type of new international order they create when they end. The outcomes here can be highly uneven. For instance, the end of World War Two created a long peace in Europe (albeit at the price of a division of the continent) that lasted for more than four decades, whereas in Asia major conflicts broke out in China, Korea and Vietnam almost immediately after the war's end.

Yet before we can start to assess the significance of a war, its impact or even its necessity, we need the war to come to an end, or at least to have an identifiable and achievable end point. With modern types of warfare, however, this is becoming more difficult. Take the Global War on Terrorism as an example. In the first place, it is harder to fight an "ism", as in "terrorism", as opposed to fighting (and killing) individual terrorists. Terrorism as such is an abstraction rather than a specific target. As a former CIA director pointed out after the 2001 terrorist attacks against New York and Washington, the only way to defeat terrorism is to "drain the swamp" that produces individual terrorists. Yet in terms of healing human anger and frustrations or resolving inequalities and political repression in all its forms, this seems an almost impossible task. Second, terrorism is a global phenomenon with recruiting and support networks extending across all continents. So terrorist groups defeated in one place can quickly and easily relocate to another. We have seen this with ISIL re-establishing itself in the Sahel and West Africa as well as Afghanistan after its defeat in Syria and Iraq. We now see the prospect of Afghanistan re-emerging as both a host territory and ideological rallying point for international jihadists in the wake of the Taliban's return to power in Kabul. This has made proclamations of "mission accomplished" highly premature as we constantly chase the evolution of terrorist groups without getting ahead of them. Third, and finally, the objectives of the jihadist groups strike many security policy experts as irrational. Caliphates based on extreme interpretations of Islam and Shariah law, the systematic persecution of other branches of Islam or of non-believers in general, the commitment to the destruction of Israel or the complete rejection of western, cosmopolitan culture are all stances that make it inconceivable to negotiate with ISIL or Al Qaeda or to reach an acceptable compromise. At least with the Irish Republican Army [IRA] in Northern Ireland or the Palestine Liberation Organization [PLO] in the Middle East, meeting around the negotiating table was always an option as the middle ground existed. Yet if jihadist terrorism cannot be defeated on a global battlefield and if we cannot negotiate peace with it, then how can the Global War on Terrorism (as George W. Bush famously called it) be brought to an end? Is terrorism designed to be a form of permanent conflict with no solution and one that we need to learn to live with?

Another form of modern warfare is known as hybrid war or the grey zone. Here adversaries use modern communication and information processing technologies to undermine each other and to exacerbate divisions and social tensions. Cyber attacks and disinformation or propaganda campaigns are the hallmarks of hybrid warfare, but we have seen states like Russia also resort to aggressive intelligence operations, targeted assassinations and even chemical weapons attacks. The challenge of hybrid warfare is to determine when it is the equivalent of an armed attack and can then come into the category of traditional warfare. NATO, for instance, has declared that a cyber attack above a certain threshold of impact could trigger its Article 5 collective defence clause. Yet this is not straightforward given that the perpetrators of hybrid warfare often cover their tracks, hide behind fake identities or keep their attacks below the level that would provoke a damaging response from the victimised state (ransomware cyber attacks against western critical infrastructure companies are a case in point). The aggressor nearly always denies responsibility and himself acts as the aggrieved victim. Myriad versions of the same event related in both the mainstream media and the social media increasingly confuse public opinion, create space for all kinds of alternative explanations and conspiracy theories, and make public opinion doubt that the truth can ever be reliably established.

The value of hybrid warfare to an aggressor is that it is potentially high gain in terms of the damage and disruption it can cause (the Russian interference in the 2016 US election is a case in point) but relatively low risk in terms of the retaliation or sanctions that the attacker can expect. So hybrid campaigns are difficult to deter, especially with traditional instruments such as tanks or nuclear weapons. The targets that can be interfered with in government, the economy, business and civil society are almost endless. Meanwhile states and organisations like the EU and NATO are hard pressed to come up with a playbook of response options (such as the expulsion of diplomats or the freezing of oligarchs' assets) that could impose a cost on the hybrid attacker and ideally deter him from further such attacks. When deterrence no longer works, limiting the impact and severity of intrusions, and recovering quickly becomes the name of the game. Resilience is now the new buzzword of modern security policy.

Yet as strategists grapple with the new forms of warfare and violent or non-violent conflict, traditional wars with conventional weapons and military operations have not gone away. In Africa fighting is underway in over 20 states (the Tigray region in Ethiopia or the renewal of conflict in Western Sahara being but the latest examples). Russia has invaded Ukraine and over 14,000 have died so far in the Donbas since 2014. Armenia and Azerbaijan fought a territorial conflict over Nagorno Karabakh in 2020 in which thousands died and the death toll in Syria has now surpassed half a million. In Afghanistan the withdrawal of the foreign forces is not producing more peace but more violence as the Taliban seize their opportunity to regain power. In addition to the dead and wounded, millions have fled as refugees and displaced persons, and the United Nations now counts over 80 million uprooted persons, the highest number yet recorded. As diplomats scramble to deal with these outbreaks of conflict, trying to negotiate ceasefires, humanitarian corridors, weapons embargoes and round table talks between the conflicting parties, they can also see looming on the horizon the growing evidence of how climate change will be the driver of future conflicts by putting stresses on water, food and other resources, fuelling more competition over these ever scarcer resources, and forcing ever more people to move to safer or more sustainable environments. The UN has calculated that by the end of this decade there will be far more climate refugees than conflict refugees, although as climate change becomes a force multiplier of stress and instability this distinction may become increasingly irrelevant.

A few years back, a good deal of academic research suggested that wars were in decline and that the number of people killed in armed conflict was going down consistently from year to year. Human society was becoming more peaceful as the residual conflicts produced fewer deaths than their 20th century predecessors. Other research suggested that peacekeeping and military interventions, despite all their imperfections and the many criticisms they have attracted, were able by and large to reduce violence and prevented the majority of states subject to these outside interventions from relapsing into conflict. This overall is an encouraging picture and, as the Harvard sociologist Steven Pinker, has put it, gives us hope that we can put our trust in the "better angels of our nature". Statistically, the global decrease in violent conflict and related conflict deaths may still hold true (particularly when compared to deaths resulting internally from gang violence and criminal activity). Yet we should not be complacent. The peace agreements reached are often fragile, even where they are properly implemented. Agreements reached among elites are often not supported by civil society or particular social or ethnic groups. Instability in one country can rapidly contaminate entire regions, as we see today in the Sahel. Moreover, the ease of acquisition and low cost of highly effective modern weapons, such as armed drones, gives certain regimes the impression that they can unfreeze frozen conflicts and once again try to achieve their strategic objectives by fighting rather than in fruitless negotiations in Geneva, Vienna or New York. This was certainly the case with Azerbaijan seeking to regain its lost territory of Nagorno Karabakh last year. As China, Russia and the US rapidly modernise their conventional and nuclear armed forces and eye each other with greater suspicion, a culture of military risk among the global great powers can quickly replace a culture of restraint. According to SIPRI, global military spending at nearly US $ 1.7 trillion has never been so high, and is continuing to rise despite the Covid-19 pandemic. The prospect of devastating great power war, inconceivable after the Berlin Wall came down in 1989, is back – or at least back in the domain of the conceivable, and it is keeping strategists and diplomats awake at night.

In short, as Trotsky famously said: "you may not like war, but war likes you". Despite periods of peace and institutional and economic progress, war seems destined to remain for a long time to come an intrinsic part of the human condition, bouncing back suddenly to overwhelm and confound us. Of course, this means that strategists and diplomats have to pay more attention for picking up the early warning signals of impending conflict and to focus more on conflict prevention. Determined international actions, such as the UN preventive deployment in North Macedonia in 1993 or the threat of major sanctions or escalatory responses, can help in this regard if they are well judged and timed; and well supported by the international community at large. We can always do better at prevention. Yet it is not an exact science, and frequently a conflict breaks out in one place while our attention is engaged elsewhere. Also we need to apply greater attention, skill and effort to how wars end. This is one of the greatest challenges in international security today. Unfortunately we seem to be getting worse rather than better at achieving this, as the record of the UN Security Council in resolving conflicts such as Syria or Libya or in responding to the UN Secretary General's appeal for a Covid-19 humanitarian ceasefire has demonstrated. What we urgently need is new research, new thinking and new answers to re-invigorate our diplomatic engagement.

Fortunately at precisely the moment we need this research and new thinking so urgently, an authoritative and insightful volume written by an international group of distinguished experts has arrived on the scene. In particular, this volume grapples with the limitations of thinking from single-disciplinary perspectives and makes an attempt at creating a dialogue between different disciplinary perspectives. When read together, the essays in the volume help understand some of the complexities of war termination from the standpoints of legal scholars, strategists, moral philosophers, historians, political scientists, and practitioners thereby widening the reader's horizon of both possible problems of and solutions to war termination. This was not an easy task to achieve, and I am proud that this volume has seen the light of day in the context of that very unique research environment that is the Center of War Studies at the University of Southern Denmark (CWS). CWS was founded almost ten years ago precisely to support multidisciplinary and interdisciplinary research on war and this volume is a clear testimony to its success as well as to the need of such an endeavour.

For all these reasons this volume stands out from many recent books on the nature of modern conflict, no matter how worthy and useful these have been. It will rapidly become essential reading and a treasure trove of analysis and practical policy advice for all those diplomats, strategists, scholars and civil society representatives who have their vital part to play in perhaps this most important of all human endeavours. I commend this volume for its academic rigour but above all for its timely policy relevance.

Introduction: How do wars end?
A multidisciplinary enquiry

Chiara De Franco, Anders Engberg-Pedersen and Martin Mennecke

ABSTRACT

The cessation of military confrontations rarely coincides with the end of war. Legal and political matters continue after the last shot has been fired, civilians driven from their homes try to rebuild their houses and their lives, veterans need to adapt to their new role in civil society, and the struggle to define the history and the significance of past events only begins. In recent years, in particular, the changes in the character of contemporary warfare have created uncertainties across different disciplines about how to identify and conceptualise the end of war. It is therefore an opportune moment to examine how wars end from a multidisciplinary perspective that combines enquiries into the politics of war, the laws of war and the military and intellectual history of war. This approach enables both an understanding of how 'the end' as a concept informs the understanding of war in International Relations (IR), in international law, and in history as well as a reconsideration of the nature of scientific method in the field of war studies as such.

Introduction

How do wars end? At a glance, war is seemingly describable as a structured sequence of events typically starting with a declaration of open hostilities and culminating with the victory of one party and the defeat of another, both validated by an armistice or a peace agreement. Upon closer examination though, just as a war can start before an official declaration is issued or without it, so it can have endings of different kinds that change over time. The development of the atomic bomb, for example, presented politicians and the military with the unprecedented possibility of a cataclysmic ending, which entailed decision-making challenges that historiography has only now started bringing to light. Those events, and the following developments of the Cold War, have contributed to a widespread understanding within IR of war as a succession of events with a clear beginning and an end. However, this conception is now challenged by the changing character of war,[1] the emergence of

---

[1] See e.g. Lawrence Freedman, *The Future of War: A History*, New York: Penguin, 2017; Hew Strachan and Sibylle Scheipers (eds.), *The Changing Character of War* (Oxford: Oxford University Press 2011).

'new wars'[2] and new weapons,[3] as well as the rise of peace studies and critical security studies, which have questioned any easy distinction between war and peace.[4] Similarly, in International Law the term 'war' has long lost its traditional meaning and has been replaced by more technical terms such as 'international' and 'non-international' armed conflict. For international lawyers the different temporal dimensions of war today include also the emerging body of 'jus *post bellum*.'[5]

This volume focuses on the specific challenge of understanding not only how wars end, but also how we can talk of the end of wars in the first place. While in the eighteenth century victory could be proclaimed by winning the field of battle, how might we today conceive of the end of war given the blurry forms that the seemingly endless war on terror has morphed into? If our standard vocabularies for the description of an event seem ill-suited to characterize the phenomena at hand, which new legal or political concepts

---

[2]See the debate following the publication by Mary Kaldor, *New and Old Wars: Organized Violence in a Global Era* (Cambridge: Polity 1999). See e.g. Mats Berdal, & David M. Malone, 'Introduction', in Mats Berdal & David M. Malone (eds.), *Greed and Grievance: Economic Agendas in Civil Wars* (Boulder, CO: Lynne Rienner 2000), 1–15; Mark Duffield, *Global Governance and the New Wars: The Merging of Development and Security* (London: Zed 2001); Edward Newman, 'The "New Wars" Debate: A Historical Perspective Is Needed.' *Security Dialogue*, 35/2 (2004), 173–189.

[3]See e.g. Christopher Coker, *The Future of War: The Re-Enchantment of War in the Twenty-First Century*, Oxford: Blackwell, 2004 and *Future War*, Cambridge: Polity Press, 2015; M. L. Cummings, 'Artificial Intelligence and the Future of Warfare', London: Chatham House, 2017, available at www.chathamhouse.org/sites/default/files/publications/research/2017-01-26-artificial-intelligence-future-warfare-cummings-final.pdf, accessed 23 October 2018; John Kaag and Sarah Kreps, *Drone Warfare*, Cambridge: Polity, 2014; Frank Sauer & Niklas Schörnig, 'Killer drones: The 'silver bullet' of democratic warfare?', *Security Dialogue*, 43/4 (2012), 363–380; Joshi Shashank, 'Army of none: autonomous weapons and the future of war', *International Affairs*, 94/5 (2018), 1176–1177; P. W. Singer, *Wired for War: The Robotics Revolution and Conflict in the 21st Century*, New York: Penguin, 2010.

[4]See e.g. Resat Bayer, 'Peace transitions and democracy', *Journal of Peace Research* 47/5 (2010), 535–546; Eric M. Blanchard, 'Gender, international relations, and the development of feminist security theory', *Signs* 28/4 (2003), 1289–1312; Barry Buzan, Ole Wæver, Jaap de Wilde, *Security: A New Framework for Analysis* (Boulder, CO: Lynne Rienner 1998); Berry Buzan and Lene Hansen, *The Evolution of International Security Studies* (Cambridge: Cambridge University Press, 2009); Berenice A. Carroll, 'Peace research – Cult of power', *Journal of Conflict Resolution* 16/4 (1972), 585–616; Catia Confortini, 'Galtung, violence, and gender: The case for a peace studies/feminism alliance', *Peace & Change* 31/3 (2006), 333–367; David Fabbro, 'Peaceful societies: An introduction', *Journal of Peace Research* 15/1 (1978), 67–83; Johan Galtung, 'Violence, peace and peace research', *Journal of Peace Research* 6/3 (1969), 167–191; Nils Petter Gleditsch, Jonas Nordkvelle and Håvard Strand, 'Peace research – Just the study of war?' *Journal of Peace Research*, 51/2 (2014), 145–158; Keith Krause and Michael C. Williams, *Critical Security Studies: Concepts and Cases* (London: UCL Press 1997); Jeff Huysmans, 'Security? What do you mean?', *European Journal of International Relations*, 4 (1998), 226–255.

[5]See only Carsten Stahn and Jann K. Kleffner (eds.), *Jus Post Bellum: Towards a Law of Transition From Conflict to Peace* (Den Haag: TMC Asser 2008), and Carsten Stahn, Jennifer S. Easterday and Jens Iverson (eds.), *Jus Post Bellum: Mapping the Normative Foundations* (Oxford: Oxford University Press 2014).

do we need to develop to understand how contemporary wars end? 'The end' as concept is surrounded by a whole cluster of related terms and denotes such varying ideas as cessation, teleology, finality, conclusion, exhaustion, etc. Uncertainties about how to identify and conceptualize the end of war raise a whole constellation of questions that are relevant to different disciplines. How do philosophers define ethical responsibilities *in bello* and *post bellum* if the boundary between war and peace is ever so blurred? How do strategists define their objectives if the teleology of action becomes uncertain? How do historians bracket the known endings of war and delve into the arguments that preceded them? Which answers can international law provide for the ending of wars – and which challenges remain or have recently arisen?

This volume combines enquiries into the politics of war, the laws of war and the military and intellectual history of war to understand how 'the end' as a concept informs the understanding of war in IR, international law, and history. It builds on research presented for the first time at a conference hosted by the Center for War Studies of the University of Southern Denmark in the autumn of 2016. Like that conference, this special issue seeks to show the added value of a multidisciplinary approach to the study of war in the 'War Studies' tradition. Initially dominated by military scholars, the field of war studies is today a genuinely multidisciplinary endeavour that 'embarks on the study of war widely and pragmatically'[6] and seeks to be relevant to academic and policy communities alike. This also reflects the fact that war itself is an area of multidisciplinarity from the point of view of scientific application (probably one of the oldest from an historical point of view), and warfare itself has been a key trigger of interdisciplinarity as "use-inspired basic research."[7] And yet, publications building on the multidisciplinarity of both warfare and the war studies tradition are still sparse. The purpose of this volume is threefold: first, it aims at being relevant to scholars from a number of different disciplines in the social sciences and the humanities and to contribute to academic debates about war termination. Second, it seeks to reach out and be relevant to practitioners. Third, the volume tries to create a platform for further communication and dialogue across disciplines gesturing towards not just a multidiscliplinary but a truly interdisciplinary approach to war studies. We have attempted to achieve the first goal by bringing together scholars from the disciplines of international relations, public international law, and the humanities and asking them to use the figure of the ending as a focal point to crystallize their contribution. At the same time, we have constituted ourselves as a multidisciplinary editorial

---

[6]Sten Rynning, 'War Studies – En Introduktion', *Oekonomi og Politik* 90/1 (2017), 3–10. Translated by the authors.
[7]Steve Fuller, 'The Military-Industrial Route to Interdisciplinarity', in *The Oxford Handbook of Interdisciplinarity* (Oxford University Press 2017), 01–26.

team, which includes an editor wearing two hats as both legal scholar and academic adviser to the Danish Foreign Ministry (Martin Mennecke). The link to practice is also sought out through a good balance between theory- and practice-oriented contributions and the inclusion of an essay on the seemingly interminable war in Afghanistan written by a former career Army officer and adviser to the US State Department (Christopher Kolenda). Finally, we have experimented with the third goal by including some reflections on the utility and feasibility of multidisciplinarity and interdisciplinarity in this introduction. The overall result is a hopefully engaging collection of essays offering new perspectives on how we can understand, manage, enforce and recollect the end of wars.

On the added value of multidisciplinarity and interdisciplinarity

Since its first institutionalization in schools and universities, scientific progress has developed within the boundaries of disciplines, which can be defined as specific bodies of knowledge or skills that can be taught and learned.[8] It was mainly in the nineteenth century, however, that urbanisation and industrialisation created clear demands on institutions to produce knowledge that could serve the growing capitalist economy and develop disciplinary expertise.[9] This process together with the emergence of the 'human sciences' as separated from the natural sciences, produced a fragmentation and specialisation of scientific knowledge, which already back then was a problem for many. In particular, starting from the assumption that there is no knowledge except scientific knowledge, since the beginning of the twentieth century positivist scholars have stressed the need to progressively unify science in order to understand the harmony and the interconnectedness of different parts of the world.[10] This, however, entailed attributing the natural sciences with some scientific superiority and a demand for all other sciences to adopt and adapt to the scientific method, as if there was only one scientific reason. A wave of contestation of such an approach to science has followed in the form of post-positivism,[11] which in the social sciences also translated into the emergence of critical, constructivist, post-modern and reflexive approaches within each discipline.

It is in this context that calls for multidisciplinary, interdisciplinary and transdisciplinary research have been made from various corners of academia

---

[8]David Alvargonzález, 'Multidisciplinarity, Interdisciplinarity, Transdisciplinarity, and the Sciences', *International Studies in the Philosophy of Science*, 25/4 (2011), 387-403.

[9]Robert Frodeman, 'Interdisciplinarity', *Encyclopedia of Philosophy and the Social Sciences*, 1 (2013), 495–497.

[10]See e.g. I. Hacking, *Representing and intervening* (Cambridge: Cambridge University Press 1983).

[11]See e.g. J. Dupre, *The disorder of things* (Cambridge, MA: Harvard University Press 1993); D. Kellner, *Media culture: Cultural studies, identity and politics between the modern and the post- modern* (London: Routledge 1995).

on the basis of the acknowledgement that different disciplines come with different epistemological concerns and distinct methods, which are the product of specific operations, relationships, and terms within each field. While transdisciplinarity demands transcending the disciplines, going across, through and beyond each individual discipline, the purpose of multidisciplinary and interdisciplinary research remains less revolutionary as they remain anchored to the independence and autonomy of each discipline. In particular, while interdisciplinarity analyzes, synthesizes and harmonizes links between disciplines into a coordinated and coherent whole, multidisciplinarity draws on knowledge from different disciplines but stays within the boundaries of those fields.[12] All three operations can of course be interpreted in a positivist or post-positivist way. On a positivist view, multidisciplinarity and interdisciplinarity are just steps towards a unified science as disciplines are perceived as producing a dangerous fragmentation of reality and transdisciplinarity has therefore a higher status as it aims to achieve not only the unification of sciences but also the unification of multiple heterogeneous disciplines and beliefs.[13] From a post-positivist perspective, instead, they are goals in and of themselves and meant to enhance knowledge and understanding of given problems while also producing greater awareness of disciplinary differences and identities, which are not perceived as necessarily hindering scientific progress.

Our approach to multidisciplinarity reflects the simple recognition that while transcending disciplines should not be a necessary end-game, dialogue across disciplines and cross-fertilisation should be sought out, especially when it comes to investigating complex social phenomena such as warfare which develop 'on the ground' also thanks to a multidisciplinary application of scientific knowledge. In addition, war and its termination continue to raise difficult questions in various disciplines which is why we undertook a multidisciplinary enquiry into its meaning. It has, however, been argued that 'claims of holistic expertise are always political claims' and that 'we cannot find powerful evidence that holistic approaches to enquiry improve our ability to act or to make effective decisions'.[14] Moreover, some evidence exists that multidisciplinary research does not improve publication chances since most academic journals are still discipline-based.[15] So why exactly do we believe two, three or four heads are better than one?

---

[12]B. C. K. Choi, and A. W. P. Pak, Multidisciplinarity, interdisciplinarity and transdisciplinarity in health research, services, education and policy: 1. Definitions, objectives, and evidence of effectiveness. *Clinical and Investigative Medicine* 29 (2006), 351–364.

[13]See e.g. Edgar Morin, 'Interdisciplinarité et transdisciplinarité', *Transversales Science Culture* 29 (1994), 4–8.

[14]D. Sarewitz, 'Against holism', in P. Galison and D. J. Stump (eds.) *The disunity of science* (Stanford, CA: Stanford University Press 1996), 73; emphasis in original.

[15]See e.g. Lee Sigelman, 'Are Two (or Three or Four ... or Nine) Heads Better than One? Collaboration, Multidisciplinarity, and Publishability', *PS: Political Science & Politics*, 42/3 (2009), 507–512.

First of all, following Clignet and Fertziger, multidisciplinarity and interdisciplinarity enhance scientific innovation as each one of them avoids what the authors call 'independent inventions', that is, redundant concepts or theories developed within different disciplines because of lack of awareness of each other.[16] Second, multidisciplinarity and interdisciplinarity imply the cultivation of a set of personal virtues such as open-mindedness, disciplinary modesty, and the ability to see things from different perspectives, which might lead to research trajectories that would be unthinkable otherwise.[17] Finally, multidisciplinarity and interdisciplinarity push a case-based approach – as with this special issue and its focus on war endings – that in turn leads to critical reassessments of scientific 'laws' and general principles.[18]

At the same time, we also recognize that calls for multidisciplinarity and interdisciplinarity have been used by academics 'to gesture toward conducting research that's more relevant than "normal" disciplinary knowledge, while avoiding the painful task of actually working with people outside the academy' and that therefore multidisciplinarity and interdisciplinarity bring with it the risk of strengthening a certain academic tendency 'to get caught up in inside-baseball debates.'[19] For that reason our attempt to initiate a joint discussion across disciplines and with practitioners on the endings of wars is anchored in the specific characteristics of the war studies tradition and its aspiration to create such dialogue.

Setting up a dialogue across disciplines

In its attempt to facilitate multi- and even interdisciplinary research, this volume has been put together with the explicit goal of enabling a fruitful dialogue across different disciplines as well as a pragmatic and at the same time reflective approach to the study of war. Following Klein, while multidisciplinarity involves encyclopedic, additive juxtaposition or, at most, some kind of coordination, but lacks intercommunication and the integration of disciplines, interdisciplinarity does precisely that – integrating, interacting, linking, and focusing.[20] This volume constitutes a multidisciplinary platform that sets the

---

[16]Remi Clignet, Allen Fertziger, 'Independent Inventions in the Social Sciences: A Plea for Multidisciplinarity', *Science Communication*, 11/2 (1989), 170–180.

[17]Frodeman, 'Interdisciplinarity'.

[18]Wolfgang Krohn, 'Interdisciplinary cases and disciplinary knowledge', in R. Frodeman (ed.) *The Oxford handbook of interdisciplinarity* (Oxford, England: Oxford University Press 2010), 31–38.

[19]Robert Frodeman, 'The Future of Interdisciplinarity: An Introduction to the 2nd Edition', in Robert Frodeman, Julie Thompson Klein, and Roberto C. S. Pacheco (eds.) *The Oxford handbook of interdisciplinarity, Second Edition, On-line edition* (Oxford, England: Oxford University Press 2017).

[20]Julie Thomson Klein, 'Typologies of Interdisciplinarity: The Boundary Work of Definition', in Robert Frodeman, Julie Thompson Klein, and Roberto C. S. Pacheco (eds.) *The Oxford handbook of interdisciplinarity, Second Edition, On-line edition* (Oxford, England: Oxford University Press 2017).

stage for a truly interdisciplinary integration. With this objective in mind, we first organised a conference where the contributors to this volume met and discussed their work amongst each other and with other representatives of the disciplines present in this collection. We then asked them to develop their contributions further, hoping the multidisciplinary conversation started at the Center for War Studies in Odense could lead them to new insights. Those essays have first been published as part of a special issue of the *Journal of Strategic Studies* that was titled 'How do Wars End?' and are now published again as part of this volume. They are written from specific disciplinary perspectives and reflect explicitly on how 'the end' of wars is conceptualized in particular fields. Yet, taken together, they form a multidisciplinary attempt to analyse current meanings and practices of war termination and their consequences for the understanding of war and peace. In some cases they also display a quest for interdisciplinary integration. To name just one example, Cian O'Driscoll, by choosing to focus on Just War Theory and its ambiguous relations with the notion of victory, has ventured into the traditionally interdisciplinary field of ethics,[21] and engaged with popular fiction and poetry to support his argument alongside debates in strategic studies and international law.

A few reflections on the essays in this volume

When reviewing the contributions for this volume, we were struck by the rich insights captured in the multidisciplinary conversations before us – but also witnessed that more work needs to be undertaken before we can speak of genuine interdisciplinarity. In fact, while sometimes interdisciplinarity arises smoothly from the convergence of two or more disciplines in a given field and even gives rise to new independent and sovereign disciplines – as it has happened, for example, in biochemistry, geophysics, cybernetics or the science of climate change – at other times it requires targeted, explicit, and institutionalised efforts. Clearly, in the field of war studies, multidisciplinarity is nowadays perceived as a must, but it has not yet produced any clear inter-disciplinary convergence. This introduction reflects the attempt of the editors at learning from the different essays in the collection. We do so with a view to emphasize the need for more actual multidisciplinarity in war studies and assert that such approach may lead toward an interdisciplinary research agenda.

A few key points emerge. The most obvious one is the elusiveness of the phe-nomenon at hand. As the chapters in the volume show, war termination – argu-ably one of the most important elements of strategic thought – has not only been

---

[21]Carl Mitcham and Nan Wang, 'Interdisciplinarity in Ethics', in Robert Frodeman, Julie Thompson Klein, and Roberto C. S. Pacheco (eds.) *The Oxford handbook of interdisciplinarity, Second Edition, On-line edition* (Oxford, England: Oxford University Press 2017).

difficult to achieve but also inherently been difficult to grasp conceptually. As Cian O'Driscoll (from the School of Social and Political Sciences at the University of Glasgow) discusses in his contribution, the end of war has long been a difficult subject for Just War Theory due to the ethical problems arising from the concept of victory. Perhaps this should come as no surprise if we consider the phenomenon itself. The plural form of the noun in our basic question, How do wars end? signals not just the recurrence of warfare as a singular event, but the plurality of forms that wars have. In a detailed typology, Joachim Krause (from the Institute for Security Policy at the University of Kiel) lays out the wide variety of wars and the often stark differences between them. To properly think war termination, one has to acknowledge the multiple distinct types of war whose specific features have a significant impact on the character of their endings. For strategists to devise plans of action that might successfully bring wars to a conclusion, they must first become fully cognizant of the differences between types of wars as well as between types of actual and potential war endings.

Thomas Obel Hansen from the School of Law at Ulster University then returns to the plurality of wars that was highlighted in the aforementioned essay by Krause – but Obel Hansen analyses whether and how international criminal law can accommodate this plurality when it comes to prosecuting massive human rights violations. Looking beyond the law, Obel Hansen discusses the political tensions that arise when international law establishes a duty to hold war criminals accountable both during and at the end of wars.

In the fourth chapter, Phillip O'Brien from the University of St. Andrews unearths the key discussions that preceded the at once most emphatic and problematic way to end a war – the use of the atom bomb in Hiroshima and Nagasaki in August 1945. If in hindsight it is difficult to imagine alternative endings to World War Two, O'Brien reveals that between the civilian and the military leadership, it was surprisingly the latter that voiced greater doubts about using the bomb. Among other things, military leaders were concerned about the reputation and global standing of the United States in a post-war world and later worried that while the bomb had ensured the Allied powers a complete military victory, it had failed to win the peace. If the atom bomb decision resulted in a decisive military victory, the meaning and the consequences of the bomb in the long run were anything but clear.

Several of these difficulties that beset war termination outlined in the first four chapters are borne out by recent events as detailed in the final essay of the volume. Christopher D. Kolenda, who served as a task force commander in the Afghan Kunar and Nuristan provinces, examines the military and political quagmire of what Mark Danner has labelled the 'forever war' – the American intervention in Afghanistan.[22] It is a cautionary, ongoing tale of what

---

[22]See Mark Danner. *Spiral: Trapped in the Forever War* (New York: Simon and Schuster 2016).

happens when politicians and military leaders fail to think hard about how wars should end.

If the individual essays cast new light on war endings within their respective fields, how might this light be deflected onto the discussions in adjacent fields? In other words, how might the different analyses of war termination gathered here learn from one another? Perhaps Just War theorists should indeed pay more attention to the concept of 'victory', as Cian O'Driscoll argues. But what are the consequences of thinking this notion of 'victory' into the strategic framework set up by Joachim Krause? Or how might the problems that have historically beset the notion of victory informed future political decisions to go to war and decide on their conclusion? Should accountability, in the form delineated by Thomas Obel Hansen, always already be a part of the mental framework of military strategists, and if yes, what might be done to ensure that it is? Will the increasing push for legal accountability and the arrival of new legal tools and institutions impact both the beginning and the end of wars in a way non-legal scholars still have to reflect in their own research? What is the response by other disciplines to the questions underlying Bailliet's call for a new approach to jus post bellum – what do scholars outside international law consider the meaning and value of 'positive peace'? Tackling such questions would be the next step. But first they must be posed.

Conclusion

In this introduction we have suggested that a multidisciplinary approach to the understanding of war ending can be both illustrated and fostered through multidisciplinary publications such as this volume and the special issue that preceded it, because they serve as platforms to open a stable and durable dialogue across disciplines. We have showed how the essays in this collection talk to one another and how the very act of reading them in sequence may lead to new questions about war termination as well as lay the foundation for an even more ambitious and interdisciplinary approach to the problem.

At the Center for War Studies at the University of Southern Denmark, we have taken new steps to facilitate such course of action. We have established a masters programme in International Security and Law where we experiment with co-teaching and simulation games to nourish an interdisciplinary mindset for the practitioners of the next generation and also to foster dialogue among faculty members from different disciplines. Every autumn we host an international, multidisciplinary conference as the one that triggered this special issue and we have recently established a multidisciplinary Network of War Studies gathering all war studies environments in Europe. One of the goals of the network is to provide interdisciplinary education to the PhD candidates affiliated to the Network's member institutions. It is our belief that such endeavours are necessary to develop the mindset and the mode of research that the subject requires.

Acknowledgments

We would like to thank the organisers and all the participants in the Center for War Studies' 2016 signature conference *How Do Wars End?* at the University of Southern Denmark. We are particularly grateful to Joseph A. Maiolo and Thomas G. Mahken for their curiosity and interest in publishing a special issue of the *Journal of Strategic Studies* that adopted a somewhat unusual approach to the study of war termination and was essential to the publication of this edited volume.

References

Alvargonzález, David, 'Multidisciplinarity, Interdisciplinarity, Transdisciplinarity, and the Sciences', *International Studies in the Philosophy of Science* 25/4 (2011), 387–403.

Bayer, Resat, 'Peace Transitions and Democracy', *Journal of Peace Research* 47/5 (2010), 535–546.

Berdal, Mats and David M. Malone, 'Introduction', in Mats Berdal and David M. Malone (eds.), *Greed and Grievance: Economic Agendas in Civil Wars* (Boulder: Lynne Rienner, 2000), 1–15.

Blanchard, Eric M., 'Gender, International Relations, and the Development of Feminist Security Theory', *Signs* 28/4 (2003), 1289–1312.

Buzan, Barry, Ole Wæver and Jaap de Wilde, *Security: A New Framework for Analysis* (Boulder: Lynne Rienner, 1998).

Buzan, Berry and Lene Hansen, *The Evolution of International Security Studies* (Cambridge: Cambridge University Press, 2009).

Carroll, Berenice A., 'Peace Research – Cult of Power', *Journal of Conflict Resolution* 16/4 (1972), 585–616.

Choi, Bernard C. K. and Anita W. P. Pak, 'Multidisciplinarity, Interdisciplinarity and Transdisciplinarity in Health Research, Services, Education and Policy: 1. Definitions, Objectives, and Evidence of Effectiveness', *Clinical and Investigative Medicine* 29 (2006), 351–364.

Clignet, Remi and Allen Fertziger, 'Independent Inventions in the Social Sciences: A Plea for Multidisciplinarity', *Science Communication* 11/2 (1989), 170–180.

Coker, Christopher, *The Future of War: The Re-Enchantment of War in the Twenty-First Century* (Oxford: Blackwell 2004).

Coker, Christopher, *Future of War* (Cambridge: Polity Press 2015).

Confortini, Catia, 'Galtung, Violence, and Gender: The Case for a Peace Studies/Feminism Alliance', *Peace & Change* 31/3 (2006), 333–367.

Cummings, M. L., *Artificial Intelligence and the Future of Warfare* (London: Chatham House, 2017), available at www.chathamhouse.org/sites/default/files/publications/research/2017-01-26-artificial-intelligence-future-warfare-cummings-final.pdf

Danner, Mark, *Spiral: Trapped in the Forever War* (New York: Simon and Schuster 2016).

Duffield, Mark, *Global Governance and the New Wars: The Merging of Development and Security* (London: Zed 2001).

Dupre, John, *The Disorder of Things* (Cambridge, MA: Harvard University Press 1993).

Fabbro, David, 'Peaceful Societies: An Introduction', *Journal of Conflict Resolution* 15/1 (1978), 67–83.

Freedman, Lawrence, *The Future of War: A History* (New York: Penguin 2017).

Frodeman, Robert, 'Interdisciplinarity', in Byron Kaldis (ed.), *Encyclopedia of Philosophy and the Social Sciences* (Thousand Oaks: Sage 2013), vol. 1, 495–497.

Frodeman, Robert, 'The Future of Interdisciplinarity: An Introduction to the 2nd Edition', in Robert Frodeman, Julie Thompson Klein, and Roberto C. S. Pacheco (eds.), *The Oxford Handbook of Interdisciplinarity*, 2nd edition, online edition (Oxford: Oxford University Press 2017).

Fuller, Steve, 'The Military-Industrial Route to Interdisciplinarity', in Steve Fuller (ed.), *The Oxford Handbook of Interdisciplinarity* (Oxford: Oxford University Press, 2017), 1–26.

Galtung, Johan, 'Violence, Peace and Peace Research', *Journal of Peace Research* 6/3 (1969), 167–191.

Gleditsch, Nils Petter, Jonas Nordkvelle and Håvard Strand, 'Peace Research – Just the Study of War?', *Journal of Peace Research* 51/2 (2014), 145–158.

Hacking, Ian, *'Representing and Intervening'* (Cambridge: Cambridge University Press 1983).

Huysmans, Jeff, 'Security? What do you Mean?', *European Journal of International Relations* 4 (1998), 226–255.

Kaag, John and Sarah Kreps, *Drone Warfare* (Cambridge: Polity 2014).

Kaldor, Mary, *New and Old Wars, Organized Violence in a Global Era* (Cambridge: Polity 1999).

Kellner, Douglas, *Media Culture: Cultural Studies, Identity and Politics between the Modern and the Post-modern* (London: Routledge 1995).

Klein, Julie Thomson, 'Typologies of Interdisciplinarity: The Boundary Work of Definition', in Robert Frodeman, Julie Thompson Klein, and Roberto C. S. Pacheco (eds.), *The Oxford Handbook of Interdisciplinarity*, 2nd edition, online edition (Oxford: Oxford University Press 2017).

Krause, Keith and Michael C. Williams, *Critical Security Studies: Concepts and Cases* (London: UCL Press 1997).

Krohn, Wolfgang, 'Interdisciplinary Cases and Disciplinary Knowledge', in Robert Frodeman (ed.), *The Oxford Handbook of Interdisciplinarity* (Oxford: Oxford University Press 2010), 31–38.

Mitcham, Carl and Nan Wang, 'Interdisciplinarity in Ethics', in Robert Frodeman, Julie Thompson Klein, and Roberto C. S. Pacheco (eds.), *The Oxford Handbook of Interdisciplinarity*, 2nd edition, online edition (Oxford: Oxford University Press 2017).

Morin, Edgar, 'Interdisciplinarité et Transdisciplinarité', *Transversales Science Culture* 29 (1994), 4–8.

Newman, Edward, 'The 'New Wars' Debate: A Historical Perspective Is Needed', *Security Dialogue* 35/2 (2004), 173–189.

Rynning, Sten, 'War Studies – En Introduktion', *Oekonomi og Politik* 90/1 (2017), 3–10.

Sarewitz, Daniel, 'Against Holism', in Peter Galison and David J. Stump (eds.), *The Disunity of Science* (Stanford: Stanford University Press 1996).

Sauer, Frank and Niklas Schörnig, 'Killer Drones: The 'Silver Bullet' of Democratic Warfare?', *Security Dialogue* 43/4 (2012), 363–380.

Shashank, Joshi, 'Army of None: Autonomous Weapons and the Future of War', *International Affairs* 94/5 (2018), 1176–1177.

Sigelman, Lee, 'Are Two (or Three or Four … or Nine) Heads Better than One? Collaboration, Multidisciplinarity, and Publishability', *PS: Political Science & Politics* 42/3 (2009), 507–512.

Singer, P. W., *Wired for War: The Robotics Revolution and Conflict in the 21st Century* (New York: Penguin 2010).

Stahn, Carsten and Jann K. Kleffner (eds.), *Jus Post Bellum: Towards a Law of Transition from Conflict to Peace* (Den Haag: TMC Asser 2008).

Stahn, Carsten, Jennifer S. Easterday and Jens Iverson (eds.), *Jus Post Bellum: Mapping the Normative Foundations* (Oxford: Oxford University Press 2017).

Strachan, Henry and Sibylle Scheipers (eds.), *The Changing Character of War* (Oxford: Oxford University Press 2011).

Nobody wins: The victory taboo in just war theory

Cian O'Driscoll

ABSTRACT

This article examines how scholars of the just war tradition think about the ethical dilemmas that arise in the endgame phase of modern warfare. In particular, it focuses upon their reticence to engage the idiom of 'victory'. Why, it asks, have scholars been so reluctant to talk about what it means to 'win' a just war? It contends that, while just war scholars may have good reason to be sceptical about 'victory', engaging it would grant them a more direct view of the critical potentialities, but also the limitations, of just war reasoning.

Introduction

How do wars end? There is an influential school of thought which contends that while wars can be concluded in a myriad of ways, they are highly unlikely to result in the outcome that most people will intuitively think of first—a decisive victory for one side, and an emphatic defeat for the other.[1] Scholars associated with this approach contend that wars are no longer likely to conclude with a clear winner and loser, but can instead be expected to drag on in a ragged fashion to the point where it is difficult to discern not just who won, but whether the war is even over.[2] The notion of victory, they conclude, is anachronistic and of little salience to the realities of contemporary armed conflict. Events in Iraq, Afghanistan, Libya and elsewhere lend credence to this perspective.[3] This point of view is, however, at odds with the enduring ubiquity of victory in elite discourse. While military historians and strategists dismiss the idea of victory as hopelessly jejune, political and military leaders continue to invoke it frequently and with just as much enthusiasm as ever.[4]

---

[1]The key statement of this position is arguably: Dominic Tierney, *The Right Way to Lose a War: America in an Age of Unwinnable Conflicts* (New York: Little, Brown and Company 2015).

[2]For example: Robert Mandel, 'Defining Postwar Victory', in Jan Angstrom and Isabelle Duvesteyn (eds.), *Understanding Victory and Defeat in Contemporary War* (Abingdon: Routledge 2007), 18.

[3]See: Gideon Rose, *How Wars End: Why We Always Fight the Last Battle* (New York: Simon & Schuster 2011).

[4]On the current ubiquity of victory talk in international relations: Cian O'Driscoll and Andrew R. Hom, 'Introduction', in Andrew R. Hom, Cian O'Driscoll, and Kurt Mills (eds.), *Moral Victories: The Ethics of Winning Wars* (Oxford: Oxford University Press 2017), 2–3.

My intention in this essay is to take the idea of victory seriously, and, following this, to examine how, if at all, it fits within the predominant Western framework for thinking about the rights and wrongs of warfare, the just war tradition. My argument, which I will develop over four sections, is that, although scholars of the just war tradition appear to have good reason for avoiding the idiom of victory, their failure to engage with it blunts the tradition's critical edge.

Section One introduces the just war tradition. Section Two details the reluctance of just war scholars to grapple with the notion of victory. Section Three examines the reasons behind this evasion, noting certain problems that pertain to the general concept of victory, and others that are specific to its relation to the ideal of just war. Squaring the circle, Section Four proffers an argument for why, despite all the apparent problems it raises, scholars of the just war tradition should engage the concept of victory. Finally, bringing all of this together in the conclusion, I will make the case for both why scholars should never forget that *just war* is just war, and why this is a matter of some importance for anyone with an interest in War Studies.

The just war tradition

Just war is commonly associated today with Michael Walzer's classic 1977 text, *Just and Unjust Wars*.[5] This book advances three main arguments. The first, directed against those who profess that the law is silent when arms are drawn, is that war is not a realm of necessity and therefore moral anarchy.[6] Instead it is a social activity, and the actions of those who partake in it are rightly amenable to ethical scrutiny and evaluation. The second argument is that the norms that delimit the practice of warfare are crystallised in what Walzer refers to as 'just war theory'. According to Walzer, just war theory comprises two modes of judgement. 'War is always judged twice, first with reference to the reasons states have for fighting, secondly with reference to the means they adopt', he writes. 'The first kind of judgement is adjectival in character: we say that a particular war is just or unjust. The second is adverbial: we say that the war is being fought justly or unjustly.'[7] The third argument is that the norms that are embedded in just war theory, and which form the substance of both modes of judgement, require revision if they are to accord with the values attested by the signatories to the UN Charter and upon which the post-1945 international order rests.

Walzer was not, of course, the originator of just war theory. Rather, he was drawing on a tradition of inquiry that can be traced back at least as far as the

---

[5]Michael Walzer, *Just and Unjust Wars: A Moral Argument with Historical Illustrations – 5th edition* (New York: Basic Books 2015).
[6]Walzer, *Just and Unjust Wars*, 3.
[7]Walzer, *Just and Unjust* Wars, 21.

fourth century CE political theology of Saint Augustine, and which later involved such luminaries as Thomas Aquinas and Hugo Grotius, among others.[8] At the heart of this tradition is the dual conviction that while the use of military force by political communities may be justified in certain circumstances, it must be subject to moral regulation. These convictions are usually bracketed under two Latinate headings that correspond to the two modes of judgement identified by Walzer. The *jus ad bellum* comprises the categories of analysis that people consult when seeking to determine what if any conditions might justify the resort to war on the part of a community. It thus encompasses the familiar concepts of just cause, proper authority, right intention, proportionality and last resort. The *jus in bello* aggregates the precepts that limit the conduct of war. These centre on the principles of discrimination and proportionality. While there is a healthy consensus among scholars regarding the centrality of these categories of analysis, they routinely disagree over how they should be defined and weighted relative to one another.

Although the tradition was not so long ago an obscure hobbyhorse kept alive only by Catholic seminary schools and historians of international law, it has experienced a revival in recent years.[9] Academic interest in the just war tradition has mushroomed since the publication of *Just and Unjust Wars* in 1977, with an upsurge in the number of universities teaching the subject, more people writing books and articles about it and the emergence of a specialist periodical, the *Journal of Military Ethics*, dedicated to servicing this new cottage industry. Alongside this, the just war tradition has become a staple on curricula at military academies the world over, and its precepts written into military codes and strategic doctrines.[10] Most interestingly, however, its concepts and terminology have become increasingly prominent in the speeches and statements of military and political leaders on the topic of war. President Obama's Nobel Peace Prize Address is an obvious example of this phenomenon, but it is not unique to him; President George W. Bush and his predecessors were also no strangers to the just war idiom.[11]

Conspicuous by its absence

As the literature on just war has grown in recent years, so too has the range of topics it has addressed. Every nook and cranny of the various *jus ad*

---

[8]The origins of the tradition are disputed. Rory Cox, 'Expanding the History of the Just War: The Ethics of War in Ancient Egypt', *International Studies Quarterly* 61/2 (2017), 371–84. On the development of the tradition: Daniel Brunstetter and Cian O'Driscoll (eds.), *Just War Thinkers: From Cicero to the 21st Century* (Abingdon: Routledge 2017).

[9]Michael Walzer, 'The Triumph of Just War Theory (and the Dangers of Success)', in *Arguing About War* (New Haven, CT: Yale University Press 2003), 3–22.

[10]Paul Robinson, Nigel de Lee, and Don Carrick (eds.), *Ethics Education in the Military* (Burlington, VT: Ashgate 2008).

[11]See: Mark Totten, *First Strike: America, Terrorism, and Moral Tradition* (New Haven: Yale University Press 2010), 80–83.

bellum and *jus in bello* categories have been explored, and they have also been extended to address emergent issues in international relations as well as new military technologies and ways of waging war. Books and articles have recently been published on, among other things, the ethics of non-violent ways of conducting war, the use of force short of war, drone warfare, the ethics of espionage, the limits of anticipatory defence, the continuing utility of sovereignty as a baseline for moral reasoning about war, the ethical challenges posed by the use of robotics and artificial intelligence in armed conflict and the implications of new conceptions of spatiality for both *jus ad bellum* and *jus in bello* categories.[12] One topic that is conspicuous by its absence is, however, victory.

At this point, scholars familiar with just war theory may query this claim by pointing to two issues that have so far been overlooked. The first is the category of 'reasonable chance of success', which many scholars identify as an integral component of the *jus ad bellum* framework.[13] It stipulates that the use of force should not be employed, even where it is otherwise justified, in situations where it is likely to end in failure. As such it serves a 'prudential' function, obliging communities to refrain from the pursuit of just but futile causes.[14] The issue here is that the idiom of victory is subsumed within a general 'prudential calculation of the likelihood that the means used will bring the justified ends sought.'[15] A vague emphasis on utilitarian calculations cast in the generic language of success, which is neither defined nor interrogated, thus forecloses more specific considerations pertaining to the termination of war. Not only, then, does the principle of 'reasonable chance of success' tell us little about victory, it replaces it with platitudes.

The second issue requires more attention. It relates to the *jus post bellum* framework. Conceived as a stand-alone category, the *jus post bellum* is a recent addition to just war thinking. It was first proposed by Michael J. Schuck in a 1994 essay, 'When the Shooting Stops: Missing Elements in Just War Theory', published in *The Christian Century*. Disgusted by the triumphalism displayed by the United States in the wake of the 1991 Gulf War—and

---

[12]Some notable recent publications include: Michael L. Gross and Tamar Meisels (eds.), *Soft War: The Ethics of Unarmed Conflict* (Cambridge: Cambridge University Press, 2017); George Lucas, *Ethics and Cyber Warfare: The Quest for Responsible Security in the Age of Digital Warfare* (Oxford: Oxford University Press 2017); James Pattison, *Just and Unjust Alternatives to War* (Oxford: Oxford University Press 2018); and Amy Eckert, *Outsourcing War: The Just War Tradition in the Age of Military Privatization* (New York: Cornell University Press 2015). This list is indicative rather than exhaustive. It is intended to offer a sense of the wide range of topics just war scholars have tackled in recent years.

[13]Colin Gray and Keith Payne present it as 'one of the six guidelines for the use of force provided by the "just war" doctrine'. Colin S. Gray and Keith Payne, 'Victory is Possible', *Foreign Policy* 39 (1980), 16.

[14]James Turner Johnson, *Morality and Contemporary Warfare* (New Haven: Yale University Press 1999), 34.

[15]Johnson, *Morality and Contemporary Warfare*, 29.

especially the jingoistic victory parade that members of the military top brass (including General Norman Schwarzkopf) celebrated on Main Street, Disneyland—Schuck argued that just war theory, as it then stood, offered no guidance for how communities should comport themselves in the aftermath of war. He proposed that a new pole of just war reasoning, which he christened the *jus post bellum*, should be formulated to meet this need.[16]

Schuck's proposition received a warm response. A number of leading scholars, including Alex J. Bellamy, Gary Bass, Brian Orend, Larry May, and Eric Patterson, endorsed it, and set about acting upon it.[17] Victory was, quite literally, the pivot for these formulations. As May framed it, for instance, the key question for *jus post bellum* theorists is 'what difference should there be between victors and vanquished in terms of post-war responsibilities?'[18] On a similar note, Bellamy submitted that one could approach the task of *jus post bellum* theorising with either a minimalist or maximalist approach, a distinction that turns on whether one assigns minor or extensive post-war responsibilities to the victors for societies they have vanquished in combat.[19]

Victory, then, functions as a threshold for *jus post bellum* deliberations. Herein lies the rub. Victory is assumed as a point of departure rather than interrogated by *jus post bellum* theorists. It is taken as a premise rather than a substantive issue or matter for inquiry.[20] This is reflected in a line from Michael Walzer's most recent essay on the topic: 'I am going to *assume* the victory of just warriors, and ask what their responsibilities are *after* victory.'[21] The explanation for this is of course that the *jus post bellum* has not in fact evolved as a framework designed to shed light on what we might term the ethics of victory. In actual fact, it has evolved as a guide for reflecting upon the rights and duties that victors obtain *after* victory has been attained by one side over the other and the transition to peace has already begun.[22]

---

[16]Michael J. Schuck, 'When the Shooting Stops: Missing Elements in Just War Theory', *Christian Century* (26 October 1994), 982–83.

[17]Alex J. Bellamy, 'The Responsibilities of Victory: *Jus Post Bellum* and the Just War', *Review of International Studies* 34 (2008), 601–25; Gary Bass, '*Jus Post Bellum*', *Philosophy & Public Affairs* 32/4 (2004), 384–412; Brian Orend, '*Jus Post Bellum*', *Journal of Social Philosophy* 31/1 (2000), 117–37; Larry May, *After War Ends: A Philosophical Perspective* (Cambridge: Cambridge University Press 2012); Eric D. Patterson (ed.), *Ethics Beyond War's End* (Washington, DC: Georgetown University Press 2012); and Mark J. Allman and Tobias L. Winwright, *After the Smoke Clears: The Just War Tradition and Post War Justice* (Maryknoll, NY: Orbis 2010).

[18]May, *After War Ends*, 1.

[19]Bellamy, 'The Responsibilities of Victory', 602.

[20]Cian O'Driscoll, 'At All Costs and in Spite of All Terror? The Victory of Just War', *Review of International Studies* 41 (2015), 805–08. Also: Mona Fixdal, *Just Peace: How Wars Should End* (New York: Palgrave 2012), 17.

[21]Michael Walzer, 'The Aftermath of War: Reflections on *Jus Post Bellum*', in Eric D. Patterson (ed.), *Ethics Beyond War's End* (Washington, DC: Georgetown University Press 2012), 37. Emphasis added.

[22]David Rodin, 'Two Emerging Issues of *Jus Post Bellum*: War Termination and the Liabilities of Soldiers for Crimes of Aggression', in Carsten Stahn and Jan K. Kleffner (eds.), *Jus Post Bellum: Towards a Law of Transition from Conflict to Peace* (The Hague: TMC Asser Press 2008), 53–77.

There have recently been some attempts to correct for the omission of victory from the *jus post bellum* framework, but there is as yet no systematic body of work on the matter.[23]

When one sets the *jus post bellum* aside and turns to more general contemporary writings on just war, the situation is compounded. Victory is even further outside the frame. Indeed, it hardly features at all in most just war texts, even when scholars are ostensibly discussing how wars end. Walzer is a case in point. He alludes to victory on several occasions in *Just and Unjust Wars*, and while his remarks are in each case intriguing, they are also minimal and not developed. Early in the text, for instance, he observes that the imperative to achieve victory in a just war can, if unchecked, militate against an army's commitment to waging war in a just manner. As he puts it, the quest for 'moral decency in battle and victory in war' are at crossed purposes.[24] Later, he notes that the intended end of any just war must be victory, but then immediately concedes that it can be difficult to discern what this means in practical terms. 'A just war is one that is morally urgent to win, and a soldier who dies in a just war does not die in vain ... But if it is sometimes urgent to win, it is not always clear what winning is.'[25] These remarks are, to be sure, food for thought, but Walzer, regrettably, does not expand on them.

Yet in mentioning victory in any kind of substantive sense, Walzer still goes far further than most contemporary just war scholars. A review of the key works in the field reveals that standard practice is to refer to the termination of war in terms of either a temporal break—the 'ending' or cessation of war—or the values for which it is fought, typically construed in terms of a balance between justice, peace, and order. Victory seldom features in such formulations. It is conspicuous only by its absence. The opening statement of Mona Fixdal's otherwise excellent book, *Just Peace: How Wars Should End*, is typical in this regard:

> How should wars end? ... I hold that any morally acceptable outcome to a war must strike a balance between the goals of justice and of peace. The war should end in a 'better state of peace', a peace that is more just and stabler than that which held before it began.[26]

Eric Patterson similarly sets out the desired endpoint of a just war in terms of a harmony between order, justice and conciliation.[27] The point in both cases is not necessarily that their approach is wrong, nor even that they

---

[23]Four contributions stand out: Janina Dill (ed.), 'Symposium on Ending Wars', *Ethics* 125/3 (2015), 627–780; Beatrice Heuser, 'Victory, Peace, and Justice: The Neglected Trinity', *Joint Forces Quarterly* 69 (2013), 1–7; Gabriella Blum, 'The Fog of Victory', *European Journal of International Law* 24/1 (2013), 391–421; and Hom, O'Driscoll, and Mills, *Moral Victories*.

[24]Walzer, *Just and Unjust Wars*, 48; also 31–32.

[25]Walzer, *Just and Unjust Wars*, 110.

[26]Fixdal, *Just Peace*, 1. Fixdal (fn.1) attributes the phrase 'better state of peace' to Basil Liddell Hart and Michael Walzer.

[27]Eric D. Patterson, *Ending Wars Well: Order, Justice, and Conciliation in Contemporary Post-Conflict* (New Haven, CT: Yale University Press 2012).

have selected the wrong values to emphasise. It is simply that victory is nowhere to be found in either formulation. It has been deftly and effectively circumnavigated. The purpose of this article is to consider whether this is the right strategy. Towards this end, it will consider the possibly that these approaches might be enriched by bringing them into dialogue with the concept of victory.

Why is victory a bad word?

Before turning to why it would be a good thing for just war scholars to engage the concept of victory, it is first necessary to devote a few words to thinking about why they have not done so already. Why, in other words, has victory become a taboo subject for just war scholars? A brief discussion of a landmark publication in the field provides a way into this question.

Larry May's 2012 monograph, *After War Ends: A Philosophical Perspective*, is the book in question.[28] As the title suggests, *After War Ends* comprises a philosophical engagement with the *jus post bellum* component of just war reasoning. It is the most comprehensive monograph available on ethics at the end of war today and is a very impressive, high-quality piece of work. What I would like to focus on here, however, is how the text is framed. The scope of the book—and indeed of the *jus post bellum*, as introduced by May—is defined by reference to a temporal demarcation: *after* war *ends*. This approach raises several red flags. Are terms like 'before', 'during' and 'after' helpful when it comes to speaking about war? Or is their tidy sequencing out of synch with the messiness of what they purport to describe? And how could one tell when or even whether a war had truly ended? It seems to me at least that this is a very sandy bottom in which to anchor *jus post bellum* analysis.

Would it not be a better idea, one wonders, to peg *jus post bellum* analysis more firmly to the idea of victory, rather than that of the ending of war? This would have the positive effect of tying *jus post bellum* reasoning to the strategic rationale of war-fighting while also (one could hope) avoiding the kind of spongy thinking that a reliance on terms such as 'after' encourages. Scholars of the just war tradition have been very resistant to this proposal. The concept of victory, they argued, has no proper place in just war thinking. The reasons they gave for this are instructive, and can be helpfully lumped into two brackets, each of which maps onto a broader set of concerns. Together they tell us something about why just war scholars have hitherto neglected the concept of victory.

The first bracket encompassed problems that are inherent to the concept of victory itself. These problems stemmed from the fact that victory turns out to be just as spongy a concept as 'after', and equally difficult to define. Interestingly, this observation is not merely an outsider's prejudice, but is

---

[28]See fn. 17.

shared by those who have spent time studying and writing about victory. Thus Robert Mandel, the author of *The Meaning of Military Victory*, describes it as a 'fuzzy, contentious, and emotionally charged' concept, while Richard Hobbs variously describes it as an 'elusive phantom' and a 'mysterious and enticing shadow'.[29] The suggestion in both cases is that victory is an inherently nebulous concept that eludes easy definition.[30] While strategists might follow Carl von Clausewitz in insisting that victory is achieved by thwarting one's opponents in battle and imposing one's will upon them, it is not clear exactly what it signifies, or how one would identify whether or when it has been achieved in practice.[31]

The case of Iraq is an illuminating reference point. Allied forces secured a decisive military victory over the Iraqi Army in the 1991 Gulf War, but the survival of Saddam Hussein's Ba'athist regime and its subsequent refusal to accept the verdict of battle led observers to discount this victory as 'hollow'.[32] Twelve years later, when a US-led coalition invaded Iraq, ousted Hussein, and seized control of Baghdad, it appeared to many that the promise of victory, frustrated in 1991, had finally been realised. The sight of President George W. Bush aboard the U.S.S. Abraham Lincoln declaring 'Mission accomplished' supported this view.[33] Yet fighting continued for many years after, and the fate of Iraq is still uncertain today—facts that led many people to query whether victory had ever been achieved in the first place. The words of Phil Klay, a short story writer who served in Iraq during the surge, capture something of this ambiguity: 'Success was a matter of perspective. In Iraq it had to be. There was no Omaha Beach, no Vicksburg Campaign, not even an Alamo to signal a clear defeat. The closest we'd come were those toppled Saddam statues, but that was years ago.'[34]

This belies that there is no obvious way of discerning whether and when victory has been won in wars like that in Iraq.[35] There is no consensus, for

[29]Robert Mandel, *The Meaning of Military Victory* (Boulder, CO: Lynne Rienner 2006), 13; and Richard Hobbs, *The Myth of Victory: What is Victory in War?* (Boulder, CO: Westview 1979), xvi, 2.

[30]This impression is supported by the fact that attempts to define victory frequently descend into typologies that distinguish different kinds or levels of victory (e.g., tactical, strategic, political, military, etc.). See: William C. Martel, *Victory in War: Foundations of Modern Military Policy* (Cambridge: Cambridge University Press 2007); and Brian Bond, *The Pursuit of Victory: From Napoleon to Saddam Hussein* (Oxford: Oxford University Press 1996).

[31]Carl von Clausewitz, *On War*, ed. by Michael Howard and Peter Paret (Princeton: Princeton University Press 1989), 1.

[32]Jeffrey Record, *Hollow Victory: A Contrary View of the Gulf War* (Washington, DC: Brassey's 1993).

[33]President George W. Bush, 'President Bush Announces Major Combat Operations in Iraq Have Ended, 1 May 2003'. Available at: https://georgewbush-whitehouse.archives.gov/news/releases/2003/05/20030501-15.html. Accessed: 16 July 2017.

[34]Phil Klay, *Redeployment* (New York: Penguin, 2014), p. 77.

[35]On this: Dominic P. Johnson and Dominic Tierney, *Failing to Win: Perceptions of Victory and Defeat in International Politics* (Cambridge MA: Harvard University Press, 2006); Jan Angstrom, 'The United States Perspective on Victory in the War on Terrorism', in Jan Angstrom and Isabelle Duvesteyn (eds.), *Understanding Victory and Defeat in Contemporary War* (Abingdon, Routledge, 2007): 94–113; and General Tommy Franks, 'The Meaning of Victory: A Conversation with Tommy Franks', *The National Interest* 86 (November 2006): 8.

instance, over the right metrics to employ in this kind of determination. At various times in the past, armies have used different markers to gauge whether or not victory had been won. Greek *poleis* equated victory with driving the enemy army from the battlefield, while later societies associated it with, among other things, the annexation of territory, the capture of the adversary's capital city, a superior body-count, regime change and the winning of hearts and minds.[36] The point to glean from this is that victory is just as difficult to identify in concrete terms as it is to define in the abstract.

Going beyond this, some scholars claim that the problem is not simply that victory is difficult to define or identify in practice, it is that is not a realistic outcome in contemporary armed conflict. The nature of modern warfare is not amenable to clear-cut endings, they argue, but instead tends to produce drawn-out endgames.[37] Russell Weigley is the main exponent of this view. He contends that, except for the period bookended by the battles of Breitenfeld and Waterloo, wars have seldom generated emphatic victories.[38] Rather, they grind on without anybody ever actually winning. To paraphrase Officer Pryzbylewski from the critically acclaimed HBO television series, *The Wire*: Nobody wins, it's just that one side loses more slowly than the other. The development of modern ways of waging war, and the advent of the War on Terror, have compounded this problem.[39] It is hard to win a decisive victory when wars are configured in such a way that they lack a clearly demarcated battlefield and are structured, not around the quest for pitched combat, but around strategic doctrines like the Israeli Defence Force's commitment to 'mowing the lawn'.[40]

The second bracket included problems pertaining to the appropriateness of using the term victory in relation to just war. The concern here was that victory is a retrograde concept that evokes forms of triumphalism and adversarialism that are at odds with the ethos of the just war tradition. According to this way of thinking, appeals to victory engender a crude win-at-all-costs mentality that undercuts the premium just war thinking places on moderation and restraint.[41] Prime Minister Winston

---

[36]On this point: Leo J. Blanken, Hy Rothstein, and Jason J. Lepore (eds.), *Assessing War: The Challenge of Measuring Success and Failure* (Washington DC: Georgetown University Press, 2015).

[37]Tierney, *The Right Way to Lose a War*; Michael Mandelbaum, *Mission Failure: America and the World in the Post-Cold War Era* (Oxford: Oxford University Press, 2016).

[38]Russell F. Weigley, *The Age of Battles: The Quest for Decisive Warfare from Breitenfeld to Waterloo* (London: Pimlico, 1991).

[39]As General David Petraues remarked of the global war against Al Qaeda, 'this is not the sort of struggle where you take a hill, plant the flag, and go home with a victory parade.' Mark Tran, 'General David Petraeus Warns of Long Struggle Ahead for US in Iraq', *Guardian*, 11 September 2008. Available at: https://www.theguardian.com/world/2008/sep/11/iraq.usa. Accessed: 16 July 2017.

[40]Efraim Inbar and Eitan Shamir, '"Mowing the Grass": Israel's Strategy for Protracted Intractable Conflict', *Journal of Strategic Studies* 37/1 (2014), 65–90.

[41]On this point: Walzer, *Just and Unjust Wars*, 47–48; also: Augustine, *City of God against the Pagans*, ed. by R. W. Dyson (Cambridge: Cambridge University Press 1998), 109–13; 118–23. For analysis: Philip Wynn, *Augustine on War and Military Service* (Minneapolis: Fortress Press 2013), 265–77.

Churchill provided an example of this in a speech on Allied war objectives delivered to the House of Commons in May 1940: 'What is our
aim? I can answer in one word: victory – victory, victory at all costs,
victory in spite of all terror; victory however long and hard the road
may be; for without victory, there is no survival.'[42] The situation in
which Churchill's United Kingdom found itself was not unique. It is
not unusual for societies waging ostensibly just wars to find themselves
between a rock and hard place, whereby they and their just cause stand
to be defeated unless they abandon customary *jus in bello* restraints
and learn to fight dirty. The choice in such instances is between winning a just war and waging it justly; it is difficult, it seems, to do both.[43]

References to victory also, it is claimed, appear callous and insensi-
tive when applied to modern war. How can it be proper to speak about
winning, the argument goes, when the issue at hand is mechanised
slaughter? As Kenneth Waltz put it: 'Asking who won a given war,
someone has said, is like asking who won the San Francisco earthquake.
... In wars there is no victory but only varying degrees of defeat.'[44]
Aristide Briand similarly observed: 'In modern war there is no victor.
Defeat reaches out its heavy hand to the uttermost corners of the earth,
and lays its burdens on victor and vanquished alike.'[45] Indeed, even
Churchill, who (see above) had committed the UK to pursue victory at
all costs, later conceded that the outcome of the war exposed the
hollowness of such talk: 'Both sides, victors, and vanquished, were
ruined.'[46] The inference in each case is that modern war, whether just
or unjust, is sufficiently ghastly in terms of its destructiveness that the
prospect of ever 'winning' it rings hollow.[47] As it is articulated in *Spoils*,
a recent novel that tackles the 2003 Iraq War, 'even if you win, you
lose.'[48] It was precisely this point that Michael Schuck, provoked by the
sight of generals celebrating the tragedy of the Gulf War as if it were a
triumph, wished to make when he initially proposed the *jus post bellum*
framework.

---

[42]Quoted in: Bond, *The Pursuit of Victory*, 142.
[43]Andrew Fiala has expressed this idea in very simple terms: 'There is an irresolvable tension between
the demands of morality and the need to win.' Andrew Fiala, *The Just War Myth: The Moral Illusions of
War* (Lanham, MD: Rowman & Littlefield 2008), 6.
[44]Kenneth Waltz, *Man, the State and War: A Theoretical Analysis* (New York: Columbia University Press
2001), 1.
[45]Quoted in: Hobbs, *The Myth of Victory*, 477.
[46]Winston Churchill, *The World Crisis, Volume V: The Unknown War* (London: Bloomsbury 2015), 1.
[47]There is an interview with Bao Ninh, a veteran of the North Vietnamese Army, in the 2017 Ken Burns
and Lynn Novick documentary, *The Vietnam War*, that bears reference here. 'People sing about
victory ... They're wrong. Who won and who lost is not a question. In war, no one wins or loses.
There is only destruction.' Quoted in: Christopher J. Finlay, *Is Just War Possible?* (Cambridge: Polity
2018), 55.
[48]Brian van Reet, *Spoils* (London: Jonathan Cape 2017), 199.

No substitute for victory?

The arguments against victory are, it would appear, compelling. They force one to think twice about both the utility and aptness of speaking about modern war in terms of 'winning'. Yet I think there is also much that these arguments miss. I want to focus here on three important and closely related matters that, when taken seriously, put a very different complexion on matters.

The first matter bears on the role that victory plays in how we typically tend to think about war. While international lawyers may prefer to address war in terms of different sequential stages (i.e., before, during and after) that correspond to different bodies of law, pundits and practitioners alike commonly associate it with the quest for victory. As General Douglas MacArthur famously put it, 'War's very object is victory ... In war there is no substitute for victory.'[49] Variations on this theme have been sounded by figures from Aristotle and Sun Tzu to Presidents George W. Bush and Donald Trump.[50] It has also been enshrined in strategic thinking, most notably in the Powell-Weinberger doctrine.[51] War poets from Robert Southey ('After Blenheim') to Wilfred Owen ('Smile, Smile, Smile') have also echoed it, and it has even been cast in stone in the motto carved above the entrance to France's most famous military academy, Saint Cyr.[52] It is also expressed in no uncertain terms by the chief protagonist in *Spoils*, the Iraq War novel mentioned earlier. 'We should play to win', Private Cassandra, the protagonist, tells her commanding officer.[53] This matters insofar as it suggests that when proponents of just war discourse avoid the language of victory, they distance themselves from how the people who command and wage wars think about them. This has implications for the ability of those same just war scholars to speak truth to power.

---

[49]General Douglas MacArthur, 'Farewell Address to Congress, 19 April 1951'. Available at: http://www.americanrhetoric.com/speeches/douglasmacarthurfarewelladdress.htm. Accessed: 16 July 2017.

[50]Aristotle, *Nicomachean Ethics*, trans. by Harris Rackham (London: Wordsworth Classics 1996), 3; Marcus Tullius Cicero, *The Republic* and *The Laws*, trans. by Niall Rudd (Oxford: Oxford University Press 1998), 83; Sun Tzu is quoted in: Mark R. McNeilly, *Sun Tzu and the Art of Modern Warfare* (Oxford: Oxford University Press 2015), 16; Bush is quoted in: Angstrom, 'The United States Perspective on Victory', 98; and Trump is quoted in: Andrew R. Hom and Cian O'Driscoll, 'Can't Lose for Winning: Victory in the Trump Presidency', *The Disorder of Things Blogspot*, 24 January 2017. Available at: https://thedisorderofthings.com/2017/01/24/cant-lose-for-winning-victory-in-the-trump-presidency/. Accessed: 16 July 2017.

[51]Caspar Weinberger, 'The Uses of Military Power'. Available at: http://www.pbs.org/wgbh/pages/frontline/shows/military/force/weinberger.html. Accessed: 28 May 2015.

[52]John I. Alger, *The Quest for Victory: The History of the Principles of War* (Westport, CT: Greenwood Press 1982), 173. The remarks of Sebastian Junger, a journalist who spent 15 months embedded with a US platoon at a remote outpost in Afghanistan, are also worth noting: 'Much of modern military tactics is geared toward maneuvering the enemy into a position where they can essentially be massacred from safety. It sounds dishonourable only if you imagine that modern war is about honour; it's not. It's about winning'. Sebastian Junger, *War* (New York: Twelve 2011), 140.

[53]Van Reet, *Spoils*, 42.

There is also a case to be a made that some level of engagement with the language of victory might remind scholars that '*just war* is also just war'.[54] Put more plainly, it would check the growing tendency to sanitise just war by speaking about it in terms that obscure its brutish realities.[55] Just wars are strategically directed violent enterprises that produce winners and losers; it would not be a bad thing for people to remember this. Not only would it refresh them of the stakes, it would also perhaps have the additional benefit of encouraging just war scholars to factor the strategic dimensions of warfare into their deliberations. So even if it is beset with difficulties, there are good pragmatic reasons why just war scholars should engage the language of victory.

The second matter is really an extension of the first. It speaks to the role that the concept of victory can play in focusing just war scholars on the utility of warfare, that is to say, on the good that it stands to accomplish in any given instance offset against the blood and treasure it is likely to cost. The words of two very different thinkers on war provide guidance here. We start with Clausewitz, who declared that no one ought to start a war without first being clear about what victory would comprise and how it could be obtained.[56] And we move from there to the sixteenth century humanist, Erasmus of Rotterdam, and in particular the advice he offered to Christian princes tasked with deciding whether or not to take their kingdoms to war. 'Let him apply just a little reason to the problem', Erasmus counselled, by 'counting up the true cost of the war and deciding whether the object he seeks to achieve by it is worth that much, even if he were certain of victory, which does not always favour even the best of causes.'[57] Combining these sentiments, we arrive at the idea that victory can serve as a focal concept for thinking about what exactly we hope or expect a just war to deliver—and at what price.

The third matter reverses the lens by providing a framework for scholars to think, not just about the costs of victory, but about the rewards that properly follow from it. The question that arises here is: What rights or legal consequences does (or should) victory in war generate? Medieval writers such as Gratian provide direction here. Gratian and his successors assumed a distinction between just and unjust wars in respect of this matter. As James Brundage explains, they held that 'Property captured and appropriated in a just war was rightfully taken and ... title to it legitimately passed to the victor. ... Rights to tangible property and intangible rights might both be

---

[54]Ken Booth, 'Ten Flaws of Just Wars', *The International Journal of Human Rights* 4/3 (2000), 316–17.
[55]As Walzer has remarked, some just war scholars seem to have forgotten that the object of their inquiry is war. Walzer, *Just and Unjust Wars*, 335–36.
[56]Clausewitz, *On War*, 579.
[57]Erasmus, *The Education of a Christian Prince*, ed. by Lisa Jardine (Cambridge: Cambridge University Press 1997), 103.

won in a just war, but not in an unjust war.'[58] Later, in the seventeenth century, legal theorists such as Hugo Grotius and Samuel von Pufendorf would complicate matters even further by inquiring as to whether victory in war created *new* legal rights for the winning party or merely vindicated *pre-existing* ones.[59] By engaging the concept of victory, then, scholars may acquire greater traction on questions pertaining to the spoils of just wars, and thus open up a new approach to *jus post bellum* reasoning.

This brings me to the culmination of my argument. Bringing the concept of victory into play re-connects just war reasoning to its own critical edge by forcing scholars to be 'honest in their just war thinking'.[60] It does so by compelling these scholars to grapple with the question of what the recourse to war can do for a given society in any particular case and whether it is ever worth the misery it sews. Pitched in these terms, rather than in the gentler discourse of justice and peace (as per Fixdal), or peace, justice and reconciliation (as per Patterson), the gravity of embarking on a just war becomes fully apparent. The scholar, if he or she is so minded, may trial this approach out by asking which if any just wars delivered an incontrovertible victory that was both worthy of the name and worth the toll it exacted. Very few wars, I wager, would pass this test. Accordingly, rather than surrendering just war thinking to a retrograde logic, as critics have suggested, bringing the concept victory more firmly into play would lend it greater stringency and thereby give it a sharper critical edge.

It might of course be objected that this is not an especially radical proposal. There is some truth to this. Even setting aside the debate about 'reasonable chance of success' canvased in Part Two of this article, the principle of *jus ad bellum* proportionality ostensibly already covers the role earmarked here for victory. The argument here, however, is that if we are serious about proportionality, we must connect it not just to conversations about body-counts or the defence of values, or to effecting a balance between good and harm, but to discussions of the kind of victory we wish to see achieved and their limitations.[61] This would prevent just war scholars

---

[58]James A. Brundage, 'Holy War and Medieval Lawyers', in Thomas Patrick Murphy (ed.), *The Holy War* (Columbus, OH: Ohio State University Press 1976), 109.

[59]See: Stephen C. Neff, *War and the Law of Nations: A General History* (Cambridge: Cambridge University Press 2005), 137–40. Also: Sharon Korman, *The Right of Conquest* (Oxford: Clarendon Press 1996).

[60]John Howard Yoder, *When War is Unjust: Being Honest in Just War Thinking – Revised Edition* (Maryknoll, NY: Orbis 1996).

[61]The literature on proportionality mostly ignores the issue of victory. For example, the key paper in the field does not make a single reference to the notion of victory or winning. Thomas Hurka, 'Proportionality in the Morality of War', *Philosophy & Public Affairs* 33/1 (2005), 34–46. A later paper by Gary Brown only notes that 'Proportionality does not preclude waging war to win.' Gary D. Brown, 'Proportionality and Just War', *Journal of Military Ethics* 2/3 (2003), 173. There are of course exceptions. Michael Walzer's statement on proportionality as it applied to the 2008–09 Gaza War is a case in point. Michael Walzer, 'The Gaza War and Proportionality', *Dissent Magazine*, 8 January 2009. Available at: https://www.dissentmagazine.org/online_articles/the-gaza-war-and-proportionality. Accessed: 16 July 2017.

from falling into the trap of treating war as a bloodless abstraction and remind them of exactly what is at stake. A stanza from the aforementioned poem by Wilfred Owen, 'Smile, Smile, Smile', illuminates this point: 'Peace would do wrong to our undying dead/The sons we offered might regret they died/If we got nothing lasting in their stead./We must all be solidly indemnified./Though all be worthy Victory which all bought.'[62]

An alternative objection is that the motive behind factoring the concept of victory explicitly into moral reflection on war is to undercut the very idea of just war. By highlighting the costs associated with victory and under-scoring the probability that any 'victory' achieved will likely be partial at best, it is true that this approach is more rather than less stringent when it comes to sanctioning the recourse to war. This, one might respond, is exactly how it ought to be. But—and it is important to be clear on this point—this approach does not preclude the possibility of just wars. It acknowledges that war may be justified in certain circumstances. It should not, then, be discounted as a form of pacifism-by-the-back-door. Its aim is not to endorse the kind of reasoning that James Turner Johnson has pillor-ied under the pejorative label *jus contra bellum*.[63] Rather it is to encourage just war scholars to bear in mind both the limits of what can be achieved by the use of force in any instance, and the suffering it necessarily brings. As such, it is offered as a means, not only by which just war thinking might be kept honest, but also as a contribution to protecting against what Reinhold Niebuhr described as 'the ironic tendency of virtues to turn into vices when too complacently relied upon; and of power to become vexatious if the wisdom which directs it is trusted too confidently.'[64]

Conclusion

In this essay, I have presented a simple argument that challenges the resistance of modern just war scholars to the concept of victory. While I have acknowl-edged the (good) reasons why scholars may wish to avoid the notion of victory, I also argue that engaging it would grant them a more direct view of the limitations of just war and what it can achieve in any given instance. As such, taking victory into account would remind them of the well-worn adage that '*just war* is just war', and the importance of always remembering this when making judgements about the rights and wrongs of the use of force. There is, however, a twist here. Where most scholars who invoke the idea that '*just war* is just war' do so in a bid to discredit the whole enterprise of just war thinking, my

---

[62]Wilfred Owen, 'Smile, Smile' Smile', in *Anthem for Doomed Youth* (London: Penguin 2015), 17.
[63]James Turner Johnson, 'The Broken Tradition', *The National Interest* 45 (1996), 28. Also: Serena K. Sharma, 'The Legacy of *Jus Contra Bellum*: Echoes of Pacifism in Contemporary Just War Thought', *Journal of Military Ethics* 8/3 (2009), 217–30.
[64]Reinhold Niebuhr, *The Irony of American History* (Chicago: University of Chicago Press 2008), 133.

conclusion is that it is rather the basis upon which that enterprise properly rests.[65] For it is only by keeping in mind the base realities of the 'war' in 'just war' that just war scholars can hope to make the right calls and reach the right judgements. The irony of this is that the incorporation of victory into just war deliberations will thus render it a less rather than more triumphalist discourse, and, conversely, a more rather than less critical resource.

There is, however, another side to this argument that I have so far avoided. This is the idea that if just war scholars will find their efforts enriched by the language of victory, scholars interested in victory might similarly benefit from tapping into the bank of ideas supplied by the just war tradition. By connecting the examination of victory to the principles of justice, peace and order that animate the just war tradition, scholars in this field may discover new ways of thinking about what has hitherto been cast in terms of a relation between 'military victory' and 'political victory'.[66] This, however, is a task for another day. In the meantime, it is appropriate to conclude on the sombre note struck by events in Mosul, where the Iraqi army's victory over the Islamic State has recently turned vicious.[67] Lamentably, these dismal developments underscore both the gravity and urgency of the matters discussed in this essay.

Acknowledgments

I would like to thank the following people for their help with this paper: Sophia Dingli, Anders Engberg Pederson, Chiara De Franco, Naomi Head, Andrew Hom, Phil O'Brien and Ty Solomon.

Disclosure statement

No potential conflict of interest was reported by the author.

Funding

Research for this paper was generously supported by the ESRC [ES/L013363/1].

---

[65]This is indeed how Booth (fn. 54) invokes it.
[66]Beatrice Heuser has already proposed something of this character, but the work remains to be done. Heuser, 'The Neglected Trinity'.
[67]John Emery, '"Victory" in Mosul: Fighting Well and the Horrors of "Winning"'. At: https://www. ethicsandinternationalaffairs.org/2017/victory-mosul-fighting-well-horrors-winning/. Accessed: 20 August 2017.

Bibliography

Alger, John I., *The Quest for Victory: The History of the Principles of War* (Westport, CT: Greenwood Press 1982).

Allman, Mark J. and Tobias L. Winwright, *After the Smoke Clears: The Just War Tradition and Post War Justice* (Maryknoll, NY: Orbis 2010).

Angstrom, Jan., 'The United States Perspective on Victory in the War on Terrorism', in Jan Angstrom and Isabelle Duvesteyn (eds.), *Understanding Victory and Defeat in Contemporary War* (Abingdon: Routledge 2007), 94–113.

Aristotle, *Nicomachean Ethics*, trans. by Harris Rackham (London: Wordsworth Classics 1996).

Augustine, Saint., *City of God against the Pagans*, ed. by R. W. Dyson (Cambridge: Cambridge University Press 1998).

Bass, Gary., 'Jus Post Bellum', *Philosophy & Public Affairs* 32/4 (2004), 384–412. doi:10.1111/j.1088-4963.2004.00019.x

Bellamy, Alex J., 'The Responsibilities of Victory: Jus Post Bellum and the Just War', *Review of International Studies* 34 (2008), 601–25. doi:10.1017/S026021050800819X

Blanken, Leo J., Hy Rothstein, and Jason J. Lepore (eds.), *Assessing War: The Challenge of Measuring Success and Failure* (Washington, DC: Georgetown University Press 2015).

Blum, Gabriella., 'The Fog of Victory', *European Journal of International Law* 24/1 (2013), 391–421. doi:10.1093/ejil/cht008

Bond, Brian., *The Pursuit of Victory: From Napoleon to Saddam Hussein* (Oxford: Oxford University Press 1996).

Booth, Ken., 'Ten Flaws of Just Wars', *The International Journal of Human Rights* 4/3 (2000), 1–23. doi:10.1080/13642980008406890

Brown, Gary D., 'Proportionality and Just War', *Journal of Military Ethics* 2/3 (2003), 171–85. doi:10.1080/15027570310000667

Brundage, James A., 'Holy War and Medieval Lawyers', in Thomas Patrick Murphy (ed.), *The Holy War* (Columbus: Ohio State University Press 1976), 99–140.

Brunstetter, Daniel and Cian O'Driscoll (eds.), *Just War Thinkers: From Cicero to the 21st Century* (Abingdon: Routledge 2018).

Bush, George W. President, 'President Bush Announces Major Combat Operations in Iraq Have Ended'. 1 May 2003. Available at: https://georgewbush-whitehouse. archives.gov/news/releases/2003/05/20030501-15.html. Accessed 16 Jul. 2017.

Churchill, Winston., *The World Crisis, Volume V: The Unknown War* (London: Bloomsbury 2015).

Cicero, Marcus Tullius., *The Republic and the Laws*, trans. by Niall Rudd (Oxford: Oxford University Press 1998).

Clausewitz, Carl von., *On War*, ed. by Michael Howard and Peter Paret (Princeton: Princeton University Press 1989).

Cox, Rory., 'Expanding the History of the Just War: The Ethics of War in Ancient Egypt', *International Studies Quarterly* 61/2 (2017), 371–84. doi:10.1093/isq/sqx009

Dill, Janina (ed.), 'Symposium on Ending Wars', *Ethics* 125/3 (2015), 627–780. doi:10.1086/679529

Eckert, Amy., *Outsourcing War: The Just War Tradition in the Age of Military Privatization* (New York: Cornell University Press 2015).

Emery, John, '"Victory" in Mosul: Fighting Well and the Horrors of "Winning"'. https://www.ethicsandinternationalaffairs.org/2017/victory-mosul-fighting-well-horrors-winning/. Accessed 20 Aug. 2017.

Erasmus, *The Education of a Christian Prince*, ed. by Lisa Jardine (Cambridge: Cambridge University Press 1997).

Fiala, Andrew., *The Just War Myth: The Moral Illusions of War* (Lanham, MD: Rowman & Littlefield 2008).

Finlay, Christopher J., *Is Just War Possible?* (Cambridge: Polity 2018).

Fixdal, Mona., *Just Peace: How Wars Should End* (New York: Palgrave 2012).

Franks, Tommy General, 'The Meaning of Victory: A Conversation with Tommy Franks', *The National Interest* 86 (Nov. 2006), 8.

Gray, Colin S. and Keith Payne, 'Victory is Possible', *Foreign Policy* 39 (1980), 14–27. doi:10.2307/1148409

Gross, Michael L. and Tamar Meisels (eds.), *Soft War: The Ethics of Unarmed Conflict* (Cambridge: Cambridge University Press 2017).

Heuser, Beatrice., 'Victory, Peace, and Justice: The Neglected Trinity', *Joint Forces Quarterly* 69 (2013), 1–7.

Hobbs, Richard., *The Myth of Victory: What Is Victory in War?* (Boulder, CO: Westview 1979).

Hom, Andrew R. and Cian O'Driscoll, 'Can't Lose for Winning: Victory in the Trump Presidency', *The Disorder of Things Blogspot*, 24 Jan. 2017. Available at: https://thedisorderofthings.com/2017/01/24/cant-lose-for-winning-victory-in-the-trump-presidency/. Accessed 16 Jul. 2017.

Hurka, Thomas., 'Proportionality in the Morality of War', *Philosophy & Public Affairs* 33/1 (2005), 34–46. doi:10.1111/j.1088-4963.2005.00024.x

Inbar, Efraim and Eitan Shamir, '"Mowing the Grass": Israel's Strategy for Protracted Intractable Conflict', *Journal of Strategic Studies* 37/1 (2014), 65–90. doi:10.1080/01402390.2013.830972

Johnson, Dominic P. and Dominic Tierney, *Failing to Win: Perceptions of Victory and Defeat in International Politics* (Cambridge, MA: Harvard University Press 2006).

Johnson, James Turner, 'The Broken Tradition', *The National Interest* 45 (1996), 27–36.

Johnson, James Turner, *Morality and Contemporary Warfare* (New Haven: Yale University Press 1999).

Junger, Sebastian., *War* (New York: Twelve 2011).

Klay, Phil., *Redeployment* (New York: Penguin 2014).

Korman, Sharon., *The Right of Conquest* (Oxford: Clarendon Press 1996).

Lucas, George R., *Ethics and Cyber Warfare: The Quest for Responsible Security in the Age of Digital Warfare* (Oxford: Oxford University Press 2017).

MacArthur, Douglas General, 'Farewell Address to Congress, 19th April, 1951'. Available at: http://www.americanrhetoric.com/speeches/douglasmacarthurfarewelladdress.htm. Accessed 16 Jul. 2017.

Mandel, Robert., *The Meaning of Military Victory* (Boulder, CO: Lynne Rienner 2006).

Mandel, Robert., 'Defining Postwar Victory', in Jan Angstrom and Isabelle Duvesteyn (eds.), *Understanding Victory and Defeat in Contemporary War* (Abingdon: Routledge 2007), 13–45.

Mandelbaum, Michael., *Mission Failure: America and the World in the Post-Cold War Era* (Oxford: Oxford University Press 2016).

Martel, William C., *Victory in War: Foundations of Modern Military Policy* (Cambridge: Cambridge University Press 2007).

May, Larry., *After War Ends: A Philosophical Perspective* (Cambridge: Cambridge University Press 2012).

McNeilly, Mark R., *Sun Tzu and the Art of Modern Warfare* (Oxford: Oxford University Press 2015).

Neff, Stephen C., *War and the Law of Nations: A General History* (Cambridge: Cambridge University Press 2005).

Niebuhr, Reinhold., *The Irony of American History* (Chicago: University of Chicago Press 2008).

O'Driscoll, Cian, 'At All Costs and in Spite of All Terror? the Victory of Just War', *Review of International Studies* 41 (2015), 799–811. doi:10.1017/S0260210515000042

O'Driscoll, Cian and Andrew R. Hom, 'Introduction', in Andrew R. Hom, Cian O'Driscoll, and Kurt Mills (eds.), *Moral Victories: The Ethics of Winning Wars* (Oxford: Oxford University Press 2017), 1–14.

Orend, Brian., 'Jus Post Bellum', *Journal of Social Philosophy* 31/1 (2000), 117–37. doi:10.1111/0047-2786.00034

Owen, Wilfred., *Anthem for Doomed Youth* (London: Penguin 2015).

Patterson, Eric D., *Ending Wars Well: Order, Justice, and Conciliation in Contemporary Post-Conflict* (New Haven, CT: Yale University Press 2012).

Patterson, Eric D. (ed.), *Ethics Beyond War's End* (Washington, DC: Georgetown University Press 2012).

Pattison, James., *Just and Unjust Alternatives to War* (Oxford: Oxford University Press 2018).

Record, Jeffrey., *Hollow Victory: A Contrary View of the Gulf War* (Washington, DC: Brassey's 1993).

Reet, Brian van., *Spoils* (London: Jonathan Cape 2017).

Robinson, Paul, Nigel de Lee, and Don Carrick (eds.), *Ethics Education in the Military* (Burlington, VT: Ashgate 2008).

Rodin, David., 'Two Emerging Issues of *Jus Post Bellum*: War Termination and the Liabilities of Soldiers for Crimes of Aggression', in Carsten Stahn and Jan K. Kleffner (eds.), *Jus Post Bellum: Towards a Law of Transition from Conflict to Peace* (The Hague: TMC Asser Press 2008), 53–77.

Rose, Gideon., *How Wars End: Why We Always Fight the Last Battle* (New York: Simon & Schuster 2011).

Schuck, Michael J., 'When the Shooting Stops: Missing Elements in Just War Theory', *Christian Century*, 26 Oct. 1994, pp. 982–83.

Sharma, Serena., 'The Legacy of *Jus Contra Bellum*: Echoes of Pacifism in Contemporary Just War Thought', *Journal of Military Ethics* 8/3 (2009), 217–30. doi:10.1080/15027570903230281

Tierney, Dominic., *The Right Way to Lose a War: America in an Age of Unwinnable Conflicts* (New York: Little, Brown and Company 2015).

Totten, Mark., *First Strike: America, Terrorism, and Moral Tradition* (New Haven: Yale University Press 2010).

Tran, Mark, 'General David Petraeus Warns of Long Struggle Ahead for US in Iraq', *Guardian*, 11 Sept. 2008. Available at: https://www.theguardian.com/world/2008/sep/11/iraq.usa. Accessed 16 Jul. 2017.

Waltz, Kenneth., *Man, the State and War: A Theoretical Analysis* (New York: Columbia University Press 2001).

Walzer, Michael, (ed.), 'The Triumph of Just War Theory (And the Dangers of Success)', in *Arguing About War* (New Haven, CT: Yale University Press 2003), 3–22.

Walzer, Michael, 'The Gaza War and Proportionality', *Dissent Magazine*, 8 Jan. 2009. Available at: https://www.dissentmagazine.org/online_articles/the-gaza-war-and-proportionality. Accessed 16 Jul. 2017.

Walzer, Michael., 'The Aftermath of War: Reflections on *Jus Post Bellum*', in Eric D. Patterson (ed.), *Ethics Beyond War's End* (Washington, DC: Georgetown University Press 2012), 35–46.

Walzer, Michael, *Just and Unjust Wars: A Moral Argument with Historical Illustrations – 5th edition* (New York: Basic Books 2015).

Weigley, Russell F., *The Age of Battles: The Quest for Decisive Warfare from Breitenfeld to Waterloo* (London: Pimlico 1991).

Weinberger, Caspar, 'The Uses of Military Power'. Available at: http://www.pbs.org/ wgbh/pages/frontline/shows/military/force/weinberger.html. Accessed 28 May 2015.

Wynn, Phillip., *Augustine on War and Military Service* (Minneapolis: Fortress Press 2013).

Yoder, John Howard, *When War is Unjust: Being Honest in Just War Thinking – Revised Edition* (Maryknoll, NY: Orbis 1996).

How do wars end? A strategic perspective

Joachim Krause ⓘ

ABSTRACT
This article gives an overview of the literature on war termination both in the fields of behaviouralism social sciences and policy-oriented strategic studies. It identifies shortcomings and problems related to both lines of research. The main problem is the undifferentiated and indiscriminate use of the term 'war'. The article proposes a categorisation of wars that could form the basis for more thorough research on the topic of war termination.

Introduction

War termination has been the subject of social scientists and strategic studies scholars for decades. However, most observers agree that the current state of the art is still poor in terms of accepted knowledge and shared policy recommendations. This article starts with an overview of the literature on this subject both in the fields of behaviouralist social sciences and policy-oriented strategic studies. It identifies shortcomings and problems related to both lines of research. The main problem is the undifferentiated and indiscriminate use of the term 'war'. So long as researchers do not differentiate among different types of war, it will not be possible to address problems of war termination more comprehensively. Since there is no accepted classification of wars which is sufficiently differentiated for such a purpose, the article proposes a categorisation of wars that could form the basis for more thorough research on the topic of war termination.

War termination in social sciences and strategic studies

When the Vietnam War drew to a close in 1975, Fred Charles Iklé published a seminal book titled 'All wars must end'.[1] In it, he summed up the problems

---

[1] Iklé 1991 (revised edition).

that had made it so difficult to terminate previous wars, beginning with World War I. He then turned to World War II, which was even a greater nightmare in terms of war termination. Also analysing the Korean War and the Vietnam War, his main conclusion was that it was easier to start a war than to settle it on reasonable terms. He not only criticised authoritarian leaders such as Ludendorff, Hitler or Mussolini for failing to seek out chances for a political settlement in view of their obvious failure to reach their respective war aims but also included democratic leaders in his scathing criticism, in particular the British War Cabinet during World War I. His recommendation was that it was imperative for any reasonable political actor who was determined to go to war, to clearly define the strategic purposes and to pave the way for an early exit from war. This book has remained highly influential: Colin Powell, then US Chief of Staff, adhered to its lessons when he was preparing for Operation Desert Storm in 1990/1991.[2]

In his book, Iklé also mentioned that the subject of war termination had found relatively little attention in the academic community – at least in relation to the vast amount of literature dealing with the beginning and the conduct of wars. This underrepresentation has not fundamentally changed since then, albeit, admittedly, today there is now a much larger body of literature on the topic than one could have found in Iklé's work. Since the 1970s, besides a few historical studies, it has been left to social scientists to take up the issue.[3] Most of these studies were based on behaviouralist empirical approaches. Many of them attempted to define variables that might explain why wars ended or why certain efforts to terminate a war ultimately succeeded or failed.[4] Others applied rational choice models to the study of war diplomacy – despite Iklè's observation that under conditions of war it will be extremely difficult to devise truly rational choices.[5] Many efforts were also devoted to the study of factors that influence the termina-tion of civil wars,[6] among them studies that highlight domestic factors.[7]

In the field of policy-oriented strategic studies, war termination was an important subject as part of the nuclear deterrence debate.[8] There were also a few studies on the subject of third power or even superpower cooperation in quelling regional wars during the time of the Cold War.[9] More recently, as part of the debate on the 'Agenda for Peace', many authors have addressed

---

[2]Powell 1995, 519.
[3]Fox 1970, Massoud 1996.
[4]Klingberg 1966, Carroll 1969, Bennett and Stam 1996.
[5]Wittmann 1979, Pillar 2014.
[6]Licklider 1995, Mason and Fett 1996, Sambanis 2000, Wagner 2000, Sambanis 2001, Walter 2002, Sambanis 2004, Fearon and Laitin 2003, Chan 2003, Hegre 2004, Fearon 2004, Dixon 2009, Kreutz 2010.
[7]Halperin 1970, Bueno de Mesquita, Siverson and Woller 1992, Stanley and Sawyer 2009, Stanley 2009a, Stanley 2009b.
[8]Kissinger 1957, Halperin 1963.
[9]Holbraad 1979, Stein 1980, Regan 1996.

the issue of post-conflict rehabilitation as a means to terminate a war since ceasefire agreements cannot/will not solve the fundamental problems behind a certain conflict.[10] Yet, there has been no attempt to consolidate these different branches of research and to arrive at comparative results. As one author already put it in the late 1970s, the amount of literature on war termination is immense, but the dissatisfaction with results of research is pronounced.[11]

Not all wars are the same

One reason for these deficits is that the term 'war' is being used in a quite superficial and indiscriminate way. Not all wars are the same. World War II was fundamentally different from the Vietnam War or from the wars in Central Africa, not to mention the war between the government of Colombia and the FARC movement or the war waged against the Islamic State. Consequently, there is no point in treating all wars in the same manner and to devise generic recipes for terminating wars. However, this is exactly what most behaviouralist and rational choice analyses seem to suggest. If social sciences – and in particular strategic studies – were to arrive at research results that could be useful for political actors, it is necessary to address the subject in a more detailed fashion and to look at different types of war.

In this regard, social sciences should take similar steps as medical research did years ago when treating cancer. 'Cancer' – like 'war' – is a generic term which is unfit for scientific purposes. The term 'cancer' just denotes a group of diseases involving abnormal and malign cell growth with the potential to invade or spread to other parts of the body. There are more than 100 types of human cancer. If physicians want to treat a certain type of cancer, they have to study the specific form of cancer very closely and devise ways and means to fight it with medication, surgery, or both. There is no such thing as a generic treatment of cancer. The same is true for wars. Many different kinds of war need to be looked at with similar scrutiny as the different types of cancer. The only common denominator among the various kinds of wars is that human beings are fighting each other in a (more or less) organised manner and at a magnitude where at least 1000 people are killed within 1 year.[12] But there cannot be any general theory of how wars end.

---

[10]Doyle and Sambanis 2000, Hegre, Ellingsen, Gates and Gleditsch 2001, Paris 2004, Paris Sisk 2009, Krause and Mallory 2010, for an overview of relevant literature see also Chetail and Jütersonke 2015.
[11]Handel 1978, 51.
[12]The author follows the definition given by the Correlation of Wars project, see Sarkees 2000. An alternative counting is being used by the Upsala Conflict Data Program, which includes every armed conflict with more than 25 fatalities per year (see Pettersson and Wallensteen 2015). For the author, 25 deaths per year is too small a number in order to classify a conflict as a war.

What is war termination?

Another difficulty lies in the term 'war termination' or 'ending war'. It is too easy to call 'war termination' a coercive bargaining process by which two interacting sides commit themselves to a settlement based upon an assessment of their relative strength and credibly.[13] In history, war termination has always signified very different things, ranging from an effective ceasefire agreement to an agreement on war termination. Sometimes wars were ended by a comprehensive approach in the form of a peace treaty or a peace conference, whereas in other cases, it concluded with an overwhelming (military) victory of one side. Often, the understanding of an end to a war varies depending on from which point of view one analyses it. For most people, World War II ended on the 8th of May 1945, but for millions, it ended years later because they were held in labour camps or were subject to ethnic cleansing. The same holds true for civil wars.[14]

It might be useful for the purpose of this article to stick with a broad definition of what war termination is: in principle, any termination of violence among organised units resulting in more than 1000 fatalities per year should be covered by this term. That does not necessarily imply the end of all violence, but often it is at least the end of widespread violence.

A Clausewitzian approach is needed

Unlike behaviouralist social sciences, there is no general school of thought on how to end a war in the field of strategic studies. This is no surprise, since the basic understanding among strategic analysts is that war is the continuation of politics with other means.[15] Hence, the issue is how to revert from an unconstrained and militarised conflict to one that is regulated in a more civilised manner, or as Carl Schmitt and, more recently, Michel Foucault have put it: politics might be described as the continuation of war by other means.[16] Since the political conflicts that cause wars are so different in kind and since the dynamics of wars are so variegated, it is practically impossible to devise any generic strategic studies theory, much less any practical guidebook on how to end a war. One can devise some general principles, but it is hard to go beyond that.[17]

What is still missing in strategic studies, however, is an attempt to categorise the varieties of conflict and war pattern with the intention of approaching the issue of war termination in a more systematic manner. Such a system has

---

[13]Slantchev 2003, Stanley and Sawyer 2009.
[14]Ghobarah, Huth and Russet 2003.
[15]Clausewitz 1989, 81–83.
[16]Schmitt 1996, Foucault 2004, 15; see also Edwards 2012.
[17]Foster and Brewer 1976, Reed 1992, Legier-Topp 2009.

to be built around (A) the different types of political motivation to initiate wars or to participate in them and (B) the likely conflict dynamics and domestic policy considerations that will have an impact on whether – and by what means – a specific war can be terminated. In the scholarly literature, many authors directly or indirectly have pointed to this necessity.[18] Yet, nobody, thus far, has made a comprehensive effort to pursue this course any further. This article attempts to further the debate by suggesting a typology of different forms of armed conflicts and the concomitant war patterns and dynamics. For each conflict category, some initial reflections are being made regarding the conditions under which wars have been ended or under which they were successfully terminated. Hence, this article presents an attempt towards a new conceptual framework within which strategic studies could address the issue of war termination in a more systematic and comparative way. The author refrains from differentiating between traditional and new wars, since there is little benefit in making this distinction.[19] Although there are some forms of war, which arguably can be called historic (like the eighteenth century Cabinet Wars), it would be premature to state that we have seen the end of major wars between heavily armed industrialised nations.[20] There will be new features to warfare in the twenty-first century due to technological innovations. But the evolution in military affairs represents a lasting characteristic of the nature of warfare throughout time. Some (in particular nuclear weapons) have changed the strategic calculus of certain types of war; others have had an impact on military doctrines under conditions of specific theatres of war. However, the basic nature of the calculus to go to war remains a political one. Therefore, when looking at wars from a Clausewitzian perspective, i.e., as the continuation of politics by other means, we do not have to treat every new feature of war as a new category, but we have to be aware of the fact that some technological innovations might have an impact on the way political actors conceive of war initiation or war termination. The starting point must always be to look at the political motives that have caused political actors to resort to violence and to address the dynamic of the conflict.

The need for a more systematic approach

In the following, at least 25 different kinds of wars are identified which have occurred during the past 300 years. This typology differs from the familiar quantitative analyses such as the Correlations of War project (COW): the scholars working within COW differentiate between intra-state (civil wars),

---

[18]Stam 1996, Goemans 2000, 11–13, Bennet and Stam 2004.

[19]For such a distinction look at Münkler 2005 and Kaldor 2006; meanwhile the use of the concept 'new wars' has somewhat ceased and there are good reasons to doubt whether this concept has any major analytic value (see Jasiukenaite 2010). The debate over 'New Wars', yet, has at least provided for a growing sensitivity for the fact that there are many forms of wars, not just the 'classical' ones, see Kaldor 2013.

[20]As suggested by Tertrais 2012.

extra-state and international wars.[21] This method is suitable for quantitative analyses, yet it is insufficient for strategic analyses. The typology suggested here might not be exhaustive. Nevertheless, it could form the starting point for a debate which has been long overdue. This typology starts from the assumption that war always signifies one actor deliberately taking up arms against adversaries. It attempts to bring in some kind of systematisation into an otherwise chaotic universe of armed conflicts usually called 'wars'. The systematisation suggested here is based on empirical observation, not on a theoretical kind of classification. Given the limited space available, this article can only sketch out what war termination could mean under conditions of different types of war. The first types of war that will be discussed can be categorised as 'historical wars', while the others are more contemporary. The historical types of wars, although hypothetical, could somehow conceivably (re)emerge and, thus, cannot be excluded from this study.

(1) **Classical limitless feudal wars** are merely a historical reminiscence. But they were a common feature in Europe for centuries, in particular from the twelfth to the early eighteenth century and the collective memory of their brutal consequences is still alive. They usually were the consequence of declining feudal orders and of the interplay of various power centres vying for control of territory and prey. The Hundred Years' War between the Kingdoms of England and France, the Thirty Years' War in Central Europe, the Spanish Succession War and the Great Northern War were typical examples for this type of conflict. Their main characteristic was the sheer unlimited duration of war and the recklessness with which civilians were plundered, mistreated and killed. The basic interest of the warring parties was to gain as much control as possible over territory and cities with the intention to exploit the rural population and to tax the citizens of towns. Besides that, religious differences and factors relating to ethnic and language groups played a growing role in increasing the level of violence. These wars often developed a dynamic of their own, making it almost impossible to terminate them: war became a self-perpetuating business and enmity became so deep seated that diplomatic agreements became impossible. In most cases, war ended because one side had to give in. There were attempts to terminate these wars by negotiation. The first successful one was the Westphalia Peace Agreement from 1648. It took seven years to reach a common understanding on how to end the many wars and campaigns which were called today the Thirty Years' War. The Peace Agreement mainly consisted of detailed provisions

---

[21]Sarkees 2000.

regulating who was to control what territory and who was to decide which religion would be practised in which territory.[22] The period of endless feudal wars disappeared with the emergence of major territorial states and the rise of the mercantilist state, which, for the first time, was based on the notion that the highest priority of the state was the well-being and security of its citizens – not their exploitation.[23]

(2) **Classical cabinet wars** were a typical phenomenon of the eighteenth and early nineteenth centuries. The War of the Polish Succession, the War of the Austrian Succession, partially the Seven Years' War and the War of the Bavarian Succession belong to this category. They were characterised by limitations, both in terms of the war aims as well as of the degree of violence.[24] Unlike the Thirty Years' War, the War of the Spanish Succession or the Great Northern War, cabinet wars were fought in a way that atrocities and plundering were to be avoided. This circumstance contributed to the relative ease with which these wars could be terminated. Often the result of a few battles defined the framework for negotiations which led to ceasefires and sometimes even to a state of relative peace with some duration. The classical cabinet wars marked a first attempt to restrain the exercise of war. Their lessons were forgotten during the Napoleonic Wars, but again applied in the early nineteenth century, where wars of the great powers were mostly limited interventions in both size and duration.[25]

(3) **Wars of imperial expansion** were a typical feature of the nineteenth and early twentieth centuries, but they can also be found in the preceding centuries as well as in ancient times. They usually had domestic sources (such as overcoming domestic strife through external expansion or imperialism)[26] and led to wars between the imperial centre and the respective subject of conquest. But often two other types of wars emerged: *wars among competing imperialist powers* over how to distribute the colonial prey among each other[27] or *balance-of-power wars* where a coalition of secondary military powers joined forces in order to fight one imperial power seeking predominance. In each of these subcategories, war termination meant something different: a war of imperial or colonial

---

[22]Osiander 2001, Krasner 1995, Teschke 2003.
[23]Williams 1970, 3; see also Poggi 1978, chapter 4.
[24]Brodie 1973, 244–252, Fuller 1961, pp. 15–25, Rotenberg 1977, 12; see also Townshend 2005, 50. However, many historians doubt whether the Seven Years War can be counted as a cabinet war, it showed many more similarities with the Napoleonic Wars, see for instance Szabo 2008.
[25]Kissinger 1995, 78–102, Albrecht-Carrié 1968, Chapman 1998, 69–81, Schroeder 1962.
[26]Hobson 1902.
[27]C.f. Langer 1951, Pakenham 1992, Wesseling 1996, Bell 2014.

conquest either ended with the victory of the aggressor or with his defeat. In the case of a successful intervention, it could translate into an open insurgency ('classical people's war', see the next category) or to an extended period of colonial submission. In the case of imperialist powers competing over colonial prey, we know about a lot of successful diplomatic endeavours leading to some kind of delineation and division between Great Britain, France and other colonial powers in the nineteenth century.[28] With regard to balance-of-power wars, a plethora of diplomatic principles, formats and instruments was devised during the eighteenth and nineteenth centuries with the aim of ending (or at least channelling) violence.[29]

(4) **Traditional people's wars** are those where the population of an area, conquered by an imperialist power, takes up arms against the invader. The classical case is the Spanish insurgency against the Napoleonic occupation 1808–13,[30] but there are many cases of guerrilla wars and broader insurgencies throughout the twentieth century, most of them being anti-colonial in nature.[31] All of these wars had in common that they were fought by the peoples and by the occupying powers with a readiness to use indiscriminate violence and to commit atrocities. Such wars have proven to be extremely difficult to end. In most cases, these people's wars either ended in a complete defeat of the insurgents (as in the case of Poland and the Baltic States in the late 1940s) or the withdrawal of the occupier or colonial power (Napoleon's forces in Spain and Central Europe as well as during the anti-colonial wars of the twentieth century). In most cases, the colonial empire's withdrawal was due to domestic circumstances that made it impossible to sustain the war effort or, because external political or military pressure became too great.[32] However, more recent wars have shown that even under conditions of complete surrender or withdrawal, a high level of violence can persist. Either the victorious occupier took revenge upon the insurgents and their families or insurgents took over the country and took revenge on collaborators as well as their families and kin. Hence, war might continue within the society and might end only after decades.

(5) **National unification wars** take place when members of a nation or of an ethnic group, who are split over different territorial entities, are fighting for national unity. This is a distinct category of warfare. Wars of such nature have been rare in recent history. In nineteenth

---

[28]See for instance Förster, Mommsen and Robinson 1989.
[29]Gulik 1955, 3–93; Morgenthau 1963, 178–225.
[30]See Glover 1971, Glover 2001, Gates 2001, Esdaile 2003, Laquer 1975.
[31]See for instance Jansen and Osterhammel 2017, Clayton 2013.
[32]See for instance Smith (Tony) 1974 and 1978, Smith (Simon) 2013.

century Europe, the Italian unification wars and the German unification wars were the most noticeable cases.[33] In both instances,
smaller and medium-sized state entities were unified by a leading
power (such as Prussia) which was using military force, often with or
against an external actor (in both cases: France and Austria-Hungary)
. Therefore, it can be argued depending on how much resistance the
unification efforts meet from external powers and by the political
entities which are to be united in the new nation, such wars can
become solvable or they drag on for a considerable period.

(6) **Major international wars (world wars)** are fought among highly
developed industrialised nations with sizeable populations under
conditions of modernity. Both Word War I and World War II belong
to this category. If the East–West conflict would have turned violent,
it might also have resulted in such a war. To a certain degree, also
the Napoleonic Wars, at least in their final phase, belonged to this
type of war.[34] The salient feature of this kind of war is that the
parties have access to seemingly endless human and material
resources.[35] These classical international wars have been the most
violent wars in history with the heaviest losses in terms of human life
and damage to infrastructure (more than 70 million fatalities in both
world wars). As it turned out, it was extremely difficult to terminate
such wars.[36] While there was no serious effort by either side to end
the fighting during World War II, at least during World War I, there
were attempts to stop fighting and to initiate a political process.
These attempts remained unsuccessful. In both cases, the war was
terminated after one side had exhausted itself (World War I) or faced
total surrender (World War II). During World War II, nuclear weapons
were developed and employed by the United States in order to
terminate the war as early as possible and by lesser casualties. The
role of nuclear weapons as a means to deter the initiation of a war or
as an instrument to terminate a war had been subject to lengthy
debates within NATO (and the public) during the times of the Cold
War. The problem was how to use nuclear weapons in an initial and
limited role against an aggressor who was also in possession of
nuclear weapons and who might employ them even before the
Western Alliance could react against an invasion. Currently, it is
hard to imagine how a new total war, akin to World War I or

---

[33]For Italy see Riall 1994 and Beales and Biagini 2003, for Germany Pflanze 1979, Wawro 1996,
Showalter 2004 and Wawro 2005.
[34]See Bell 2007.
[35]See for World War I Stevenson 2004 and Broadberry and Harrison 2005, for World War II see
Weinberg 1994 and Harrison 1998.
[36]Kecskemeti 1958, Goemans 2000.

World War II, could emerge. However, such a contingency cannot be ruled out. If it was to emerge, it will be extremely difficult to terminate it. The only factor that works towards war termination is the common fear of the incalculable consequences of nuclear war.

(7) **Major international wars between less developed nations**: These are wars that have some similarities with wars among highly developed nations. Yet, one important distinction is that the warring parties might have a sizeable population but cannot sustain a lengthy war effort due to the lack of material resources and/or organisational skills. The wars between India and Pakistan fell into this category as well as the wars between Israel and Egypt in the 1960s and 1970s.[37] External supporters of the warring parties (mainly the Unites States and the Soviet Union) were crucial since only they could provide them with the means to continue fighting – if only for a few days or weeks. On the other hand, both Washington and Moscow were anxious to avoid further escalation or, worse still, to be drawn into the conflict. Thus, there have been cases in which both former superpowers cooperated in order to end the war under acceptable or at least face-saving conditions.[38] Since more and more countries of what was formerly known as the 'Third World' today have become industrialised and can be considered to be industrial threshold states, it is to be expected that future wars between such countries might show more similarities with World War I and World War II. However, it is important to note that the geopolitical conditions of Europe – where sizeable and powerful nations were stuck together on a relatively small territory, implying dangerous strategic risks and opportunities – that led to both World Wars do not apply to any region of the world today (not even in the much-discussed Asia-Pacific region).[39] The only exception so far is the Korean Peninsula. Here, war is a possibility and war termination might be extremely difficult, because of the huge forces involved on both sides. In the worst case, a war on the Korean Peninsula might only be ended by the use of nuclear weapons.

(8) **Limited wars of aggression or border disputes**. These are wars where one state (either alone or in cooperation with others) tries to conquer and annex a smaller neighbouring state, or where it attempts to take control over a territory which has been under the

---

[37]See for the wars in Middle East Bailey 1990, Hammel 1992, Gawrich 2000, Pollack 2004; for the wars in South Asia see Brown 1972 and Ganguly 1986.

[38]For a critical assessment of the role of US and Soviet Union in the Middle East, see Quandt 2005 as well as Stein 1980; from the perspective of the then US Secretary of State for the 1973 Yom Kippur war see Kissinger 1982, 450–666, see also Holbraad 1979.

[39]Krause 2014.

jurisdiction of another state. The occupation of Kuwait by Iraq in summer 1990 was the most recent case of this kind.[40] Such a war either ushers into a people's war (see category 4) or is terminated by an international military intervention that has the purpose of evicting the aggressor. A further possibility is that resistance towards the occupying force remains limited or is being put down and that the annexation is accepted by the international community (such as the conquering and annexation of Tibet by the People's Republic of China in 1950). Wars over border disputes have been subject to international diplomatic mediation efforts in the past. Whenever the disputed areas were uninhabited, a freeze of military actions was helpful. In the case of populated and strategically or symbolically important areas, ceasefire agreements usually were feasible; however, often they did not solve the problem but resulted in freezing the conflict.[41]

(9) **War over the control of islands and maritime areas**. We have not witnessed such form of conflict since the Falklands War in 1982. Such a war might remain limited as it involves maritime forces, which are both very expensive and vulnerable. The conflict, hence, could become subject to international mediation as both sides are interested in avoiding losses of capital ships and, consequently, major loss of lives. However, a conflict over a maritime area might become a focal point of major power competition if at least one side considers control over that particular area of utmost strategic importance – for whatever reason. In that case, either the other side gives in or a major showdown of naval forces (including aircraft and missiles) might arise. The conflict over the South China Sea and the Eastern China Sea might take such a direction.[42] There is no evidence so far which could give some guidance as how such a showdown might go off and into which domains it might escalate. Hence, it is hard to suggest ways and means how to terminate such a naval campaign.

(10) **Postmodern, hybrid wars**. These are wars where one country attempts to destabilise another country not only by insurgents, guerrilla fighters and classical means of subversion but also by cyber war and information war. The aim is to change territorial borders by eroding them and to keep a neighbouring state in a state of permanent intimidation. If the attacking country is larger than the attacked state, it will be hard to end such a war, unless there is an external actor ready to stand in for the security of the

---

[40]See Cordesman and Wagner 1996; Finlan 2003.
[41]See for instance Freire 2003, Croissant 1996.
[42]For East Asia see Emmers 2010.

attacked state.[43] However, it is extremely difficult to end such a war in a way that the aggressor would not gain benefits from it. Russia has been quite successful in cutting out territories from states that were former members of the Soviet Union. From Georgia they took the territories of South Ossetia and Abkhazia, from Moldova they cut off Transnistria and from the Ukraine they annexed the Crimean Peninsula and a sizeable part of the Eastern Donbas area. What makes war termination extremely difficult is the fact that the Russian government (against all evidence) pretends not to be part of the war and that it installs small political entities in the occupied territories (often composed of adventurers, criminals and extremist) which are entirely dependent on Russian support, but whose main task it is to thwart any international mediation effort. The attempts by Germany and France in the wake of Russia's hybrid invasion of the Donbas area, to mediate a ceasefire between the government in Kiev and the so-called peoples' republics, are a telling story of how complex the problems are in terminating hybrid wars.[44]

(11) **Educational wars or wars of punishment** are a category of war that has rarely occurred in recent history. The purpose of such a war is not to conquer territory or to occupy permanently another state but to 'teach an opponent a lesson'. It is a typical behaviour for a state like China, which pursues a kind of tributary approach towards its neighbouring states. The known examples were the war against India in 1962 (which began as an armed skirmish over some unpopulated mountainous areas turned into a full-fledged invasion) and the war against Vietnam in 1979, which turned out disastrously from a Chinese point of view.[45] Given the growing military strength of China, we might see more of such educational wars in the future. Both wars ended by unilateral withdrawal. External political interventions showed no impact.

(12) **Civil wars, in which governments are facing an insurgency** supported by major parts of the population, are a common feature in history. The first such known war in modern history was the German Peasants' War from 1525. The civil war in England in the seventeenth century began as a public uprising against the ruling elites, as did the French Revolution in 1789, thus putting it into this category of war. The first phase of the Russian Revolution (early 1917) is also worth mentioning in this context. The civil war in Syria is a recent example. Such wars are difficult to end. In most cases, it was the

---

[43]See Mazaar 2015.
[44]Kostanyan and Meister 2016.
[45]See Calvin 1984 for the Indo-Chinese War and Zhang 2005 for the Chinese-Vietnamese War.

complete defeat of the insurgents (as during the German Peasants' War) or the complete success of the revolution, which ended the violence. Otherwise, as violence becomes more widespread, such civil wars can turn into a lasting civil war between different armed groups/actors (see the next category) often involving external powers as well. However, even under conditions of complete failure of the insurgency or in the case of a complete success of the revolution, a high level of violence continues for considerable time. Either the government takes revenge on the insurgents and their families or insurgents take over the country and seek revenge among members of the former ruling class, their families and clans. Outside intervention might transform an insurgency into a long-lasting war, in particular if two or more competing powers are involved.

(13) **Civil wars, in which major armed forces fight each other over the political direction their country should take**. The US Civil War from 1861 to 1865, the Russian Civil War from 1917 to 1922 and the Chinese Civil War from 1945 to 1950 are the most prominent examples. They were extremely violent with countless atrocities being committed by all sides. The number of victims in China and Russia, however, was considerably higher than those of the US Civil War, i.e., in the area of 20–30 million fatalities. Ending such wars is usually impossible unless one side is defeated. Even defeat does not necessarily imply that violence ends. Often the war is continued in form of death squadrons, torture and executions in prisons, as well as inhumane treatment in labour camps.

(14) **Civil wars, in which organised violence is being exercised between members of different ethnic or religious communities (tribes)**. Such wars can take place within territorial borders but often rage across borders, in particular when these borders were drawn without paying due respect to traditional lines of demarcation and influence. Furthermore, they can erupt after a functioning state has begun to decay. Many states in African and the Greater Middle East (as well as the former Yugoslavian federation or the former Soviet Union) today have borders which were drawn arbitrarily in the past and have become a principle cause for the outbreak of civil wars. As the wars in former Yugoslavia have demonstrated, violence might proliferate to such a degree that it needs external intervention to end the carnage. Once a ceasefire agreement has been reached, often international assistance is needed in order to create a political framework within which the warring ethnic or religious group might coexist. In many cases, such political arrangements have to build upon a kind of geographic separation or even segregation within

society. Sometimes, even secession might turn out to be the most promising solution. The recent example of the decay of Libya demonstrates how quickly an entire state can collapse. Ending such civil wars is one of the most complicated tasks. It not only needs a concept for building a new political order and a functioning polity but also has to involve measures by which militias can be disarmed and disbanded. At the same time, new jobs have to be created in order to absorb former combatants.

(15) **Civil wars fought over secession**. Such wars might be easiest to prevent by a federal solution; in other words, by granting a certain degree of self-governance and autonomy within a specific territory. In case violence has already occurred, again it is the degree to which blood has been shed and atrocities have been committed that can determine whether or not a political solution is possible. If there is no place for a political settlement, either of the two sides might prevail ending the fight on its terms. This could spell secession (as in the case of Bangladesh, South Sudan, Eritrea, Kosovo or Timor-Leste) or it could mean the victory for the central state (as in the case of the Biafra War or of the Tamil Tigers' war for independence from Sri Lanka). Both secession and restoration of the rule of the central state usually will be followed by displacement of people, persecution of political activist and internal violence and suppression. Very few separatist conflicts were successfully terminated through a political solution by which the different political interests were balanced against each other.

(16) **Civil wars, where a government is fighting organised crime**. Such civil wars are taking place today in countries such as Mexico, Honduras, Costa Rica or El Salvador. To a certain degree, the fight of the Colombian government against the FARC rebels in its late phase was also a fight against a political movement, which have become indiscernible from a criminal militia. The fact that these kinds of civil wars mainly can be observed in Latin America is linked to the fact that Central America forms the main route of the narcotics trade to the United States. Thanks to the enormous profits these business promises, criminal gangs meanwhile have the means available to form powerful militias and to destabilise the state by violence, bribery and corruption. There are two possible outcomes of such wars: either the state prevails or the criminal gangs. A negotiated peace is practically impossible and not desirable, since it would legitimise criminal activities.

(17) **Civil wars with a self-perpetuating war economy**. Such civil wars are characterised by the de-statisation (privatisation) of organised violence and the emergence of a war economy that helps to perpetuate the war, which often is fed by the exploitation (so-called taxation) of the local

population through militias and by outside money and arms supplies.[46] Such dynamics usually emerge when local actors have gained control over rare earths and minerals (such as diamonds) or natural resources (gas and oil) and when they were able to find external customers who are ready to pay for these goods irrespective of their origin. The best chance to end such a war is to cut the lines of delivery between local warlords and international customers or to make it difficult for customers to do business with the raw materials and diamonds acquired by warlords (as was attempted in the Kimberley Process Certification Scheme). However, in most cases, such measures alone will not suffice. Experiences with West and Central African states have demonstrated that it also might take military interventions under UN mandates to end such a war. Usually it takes years, if not decades to terminate such conflicts and often the consequences will last for decades.

(18) There is also a category of **anti-civilisation wars**, where states belong to an expanding civilisation encounter less developed social entities (tribes), which are willing to fight the dominant civilisation. Such encounters are known from ancient history to the period of colonialisation and imperialism and often did not find their way into history books. Most anti-civilisation wars ended with the victory of the civilisation; however, often the victory did not result in a further spread of civilisation, but rather in the destruction of the traditional society, which meant local violence and disorder. Today, iterations of these encounters are less frequent and obvious but have again appeared in anti-Western militant ideologies and often form the background for terrorist attacks or guerrilla warfare.

(19) **Wars fought over religion** were one the main features of organised violence in the Middle Ages. The most prominent cases were the campaigns to spread Islam, which began in the late seventh century,[47] the Crusades (eleventh to fifteenth century)[48] and the religious wars in Central Europe during the sixteenth and seventeenth centuries. Today, the heightened role of Islam as an ideology of resistance against Western hegemony and the rivalry between Shia and Sunni within the Muslim world are indications that we might witness a revival of religious wars, or wars fought with the intention to spread a certain religion or to defend it against an assumed aggression. Terminating such wars would be an extremely difficult task. There are no precedents and there is hardly any hope that zealous religious ideologues might be able to agree to any compromise.

---

[46]Such wars are often called 'New Wars', see Münkler 2005 and Kaldor 2006.
[47]See Kennedy 2007, Donner 2014.
[48]See Nicholson 2004, Tyerman 2007, Riley-Smith 2009, Asbridge 2012.

(20) We also have to reckon that there are **wars with extremist organisations, which have taken control over territory or have begun to organise themselves as a state**. Surely, the most prominent case in kind is that of the so-called Islamic state (or caliphate), which was founded in 2014 on a territory encompassing parts of Syria and Iraq that had been brought under control by an extremist Islamist organisation. However, there are examples of such wars in other places as well, such as the seizure of Southern Lebanon by the extremist Shiite organisation Hezbollah or of the Gaza Strip by the radical Islamist movement Hamas. In all cases, these territories are being used as a basis to fight the West writ large, corrupt Arab regimes or the state of Israel in particular. So far, there have been two strategies to stop such fighting: either to form coalitions in order to liberate the respective territories from the extremist movement (such as in the case of the Islamic State) or to find a way to deter such movements by a mixture of defensive measures and punitive air strikes (as in the case of Israel dealing with Hamas and Hezbollah).[49]

(21) **Wars of annihilation** are a category of extreme warfare, which has occurred very rarely in modern history. By this term, any war is meant that has the purpose of destroying a whole society or all members of a certain state by killing the population or enslaving it. The campaign waged by the Third Reich against Poland in 1939 and later (1941) against the Soviet Union were wars of annihilation. Today, many people consider such a war impossible and unthinkable. Yet, given the constant threats by Iran against the state of Israel, it cannot be excluded anymore, in particular given the possibility of nuclear escalation.[50] Terminating such a war means to stop it before annihilation can materialise. Under conditions of nuclear threats, stopping such a war can mean to resort to a preemptive, disarming strike, although this entails very high risks.

(22) A special category is **wars fought by a coalition of states under a collective security mandate**. This is – besides the war to fend off an external aggression – the only kind of war allowed under the UN Charter. There are very few examples of such wars, the most important example being the liberation of Kuwait by an international coalition of states in 1991. The NATO air campaign against Serbia in 1995 also qualifies to be placed into this category. In any case, early war termination is usually very important, as the impression should be avoided that the war effort had different political aims.

---

[49]Cordesman, Sullivan and Sullivan 2007.
[50]See further details in Krause 2012.

(23) Similar to the aforementioned category are **limited military operations, often under collective security mandates, with the purpose
of stabilising a specific territory or state**, which has witnessed some
form of a civil unrest previous to the military intervention. As a rule of
thumb, these operations are not intended to be a war per se but rather
to end an existent civil war and to support a new state structure and the
rehabilitation of the economy.[51] As was the case in Afghanistan after
2001, such a stabilisation operation can turn into a guerrilla war when
the international force comes under attack by insurgents or terrorists. It
might even turn into an anti-civilisation war. Under such a condition,
a viable exit strategy becomes a major problem. The withdrawal of the
international force without successfully stabilising the conflict zone
would aggravate the security situation on the ground. On the other
hand, an escalation of the war could turn into an endless mission. What
might help is a counter-insurgency strategy that will focus on winning
the hearts and minds of the population.[52]

(24) The category of **regime change wars bears some resemblance to
the two preceding types of war**. Regime change wars are either
pursued by one major power unilaterally or together with a coalition
of states (which is supposed to give the intervention more legitimacy). The motivation might be benign (removing a brutal and
despotic regime which poses a threat to both its own population
as well as for international peace) or malign (a big power forcing
a smaller state into obedience); the procedures have always been
the same: a country will be occupied by foreign forces until the
government has been removed. Regime change wars are the modern equivalent of traditional wars of succession (which either
resulted in an endless feudal war or into a cabinet war). The first
modern regime change war took place in 1814, when a coalition of
European powers invaded France. The intention was to force
Napoleon to step down. The final stages of World War II are also
important to mention in this context. The removal of the Nazi
regime was the declared aim. Examples that are more recent are
the occupation of Hungary (1956) and Czechoslovakia (1968) by the
Warsaw Pact.[53] The most prominent case in recent times was the
war against the Saddam Hussein regime in Iraq (2003). Usually such
wars end with the victory of the occupier but such as in the case of
Iraq, they can also provoke an insurgency, which might later transform the campaign into a people's war.

---

[51]Schroeder (Robin) 2014.
[52]Nagl 2005.
[53]See Windsor and Roberts 1969; Györkei, Kirov and Horvath 1999.

(25) The final category mentioned here is the **preventive (disarming) wars**. These are wars led against a state (or a group of states) that is (are) thought to present a danger for international peace. The intervention should remove this threat before it really might develop. There are many wars in modern history, which were justified in this way. In most cases, the alleged intention turned out to be rather a pretext for rather familiar motives, such as imperial conquest and expansion. However, there have been wars, which were truly motivated by the intention to avoid war through an armed intervention. In March 1815, for instance, the major powers chairing the Vienna Congress set up an army to prevent Napoleon after his return from Elba from resuming war again.[54] The initiation of the Six-Days War by Israel in June 1967 might also be counted as a preventive war (almost in a preemptive mode).[55] Such wars can end within a short period of time – because this is exactly what is intended. However, they might usher in a full-fledged protracted war, in particular if the war is directed against a major industrialised state. Some authors, for instance, are concerned that China might be tempted to destroy the whole military presence of the United States in East Asia with one decisive blow in the not-too-distant future.[56]

This categorisation does not necessarily imply that every war must fall within one specific category. On the contrary, it is reasonable to assume that wars can belong to two or more categories at the same time or that a specific war starts within a certain category and subsequently transforms, thus moving into a different category. In particular, major and long-lasting wars are often a combination of different kinds of war taking place simultaneously on different theatres.

As this list might suggest, given the variegated nature of wars and their respective dynamics, it would be impossible devising a common method for terminating all wars. It would make more sense to look at these different categories in a comparative way. Consequently, one might find that some types of war might qualify to be terminated more easily while others might not. What might help ending violence in one place could be detrimental in other cases. The conceptual framework suggested here can be helpful in guiding such research.

---

[54]Jomini 1864, 54.
[55]See Oren 2002; see also Bowen 2003.
[56]See Tol, Gunzinger, Krepinevich and Thomas 2010, 20; White 2012, 74; Friedberg 2014, 82.

Concluding remarks

The real challenge, however, might be do devise concepts and ideas for terminating those wars which are currently going on: among them are the Russian hybrid insurgency in the Eastern Ukraine, the war in Syria, the war in Yemen, the Mexican war against the narcotic insurgency, the still raging counter insurgency war in Afghanistan, the civil war in South Sudan, the various local wars in Congo, the insurgencies in Somalia, Nigeria and the Western Sahara and other small wars. They all fit into one or two of those categories mentioned above. In devising concepts for terminating these wars and civil wars, the real tasks starts with applying the suggested conceptual approach and to look into the intentions of the involved actors, their use of military means, the dynamics that have emerged during the war and the opportunities for war termination as well as for the selection of strategies and instruments needed to achieve an end to war.

The big white elephant in the room, however, is the utility of nuclear weapons for the termination of wars. As Bernard Brodie had already put it in 1946, the invention of nuclear weapons entails the necessity to rethink war and the means to end it. His line of argumentation, according to which the chief purpose of our military establishment must be to avert wars,[57] has proven wrong by history insofar as many new and 'innovative' ways of war have emerged under the nuclear umbrella. During the past three decades, nuclear weapons had no role at all in bringing a local war or civil war to an end.[58] However, the pivotal role of nuclear weapons in avoiding or ending big wars might come up again. Nuclear weapons might reemerge as a major factor insofar as big power clashes cannot be excluded anymore in the not too far future.

There are two theatres, where such clashes involving nuclear weapons or at least nuclear threats might occur: the South China Sea and the Baltics. In the South China Sea (as well as in the East China Sea), China is building up an impressive missile threat against US military assets and naval forces, which has led the United States to reconsider basic strategic choices in the region.[59] Given the ongoing territorial dispute over major maritime areas and given the stakes involved, a nuclear escalation might be possible. In the Baltic region, Western observers are concerned about the growing capability of Russia to occupy Baltic Sea member states of NATO as well as neutral Sweden or Finland. The Russian build-up of a conventional and hybrid invasion capability goes along with preparations to establish a nuclear escalation potential in the region by introducing sea-based and even land-based short- and medium-range missiles and cruise missiles, which could be loaded with nuclear weapons. Under such conditions, war

---

[57]Brodie 1946, 76.
[58]Paul, Harknett and Wirtz 2000.
[59]Heginbotham et al. 2015, Kirchberger 2015, Shugart 2017.

termination becomes a different subject, since for Russia war termination would imply to stop successfully a regional war of aggression by threatening the use of nuclear weapons, for NATO it would mean how to thwart any such Russian calculus.[60]

Disclosure statement

No potential conflict of interest was reported by the author.

ORCID

Joachim Krause (iD) http://orcid.org/0000-0002-8217-5685

Bibliography

Albrecht-Carrié, René, *The Concert of Europe 1815–1914* (New York: Harper's Torchbook 1968).
Asbridge, Thomas, *The Crusades: The War for the Holy Land* (London and New York: Simon & Schuster 2012).
Bailey, Sydney, *Four Arab-Israeli Wars and the Peace Process* (London: The MacMillan Press 1990).
Beales, Derek and Eugenio Biagini, *The Risorgimento and the Unification of Italy* (London: Longman 2003).
Bell, David A., *The First Total War. Napoleon's Europe and the Birth of Warfare as We Know It* (Boston: Mariner Books 2007).
Bell, P. M. H., *France and Britain, 1900–1940: Entente and Estrangement* (Abingdon: Routledge 2014).
Bennet, D. Scott and Allan C. Stam, *The Behavioral Origins of War* (Ann Arbor: The University of Michigan Press 2004).
Bennett, D. Scott and Allan C. Stam, 'The Duration of Interstate Wars, 1816–1985', *American Political Science Review* 90/2 (1996), 239–57. doi:10.2307/2082882

---

[60]Kroenig 2018.

Bowen, Jeremy, *Six Days: How the 1967 War Shaped the Middle East* (London: Simon and Schuster 2003).

Broadberry, Stephen and Mark Harrison eds, *The Economics of World War I* (Cambridge: Cambridge University Press 2005).

Brodie, Bernard, *The Absolute Weapon: Atomic Power and the World Order* (New York, NY: Harcourt, Brace & Co. 1946).

Brodie, Bernard, *War and Politics* (London: Cassell 1973).

Bueno de Mesquita, Bruce, Siverson, Randolph M. and Woller, Gary, 'War and the Fate of Regimes: A Comparative Analysis', *American Political Science Review* 86/3 (1992), 638–46. doi:10.2307/1964127

Calvin, James Barnard, *The China-India Border War 1962* (Washington, DC: US Marine Corps Command and Staff College 1984).

Carroll, Berenice A., 'How Wars End. An Analysis of Some Current Hypotheses', *Journal of Peace Research* 6/4 (1969), 295–320. doi:10.1177/002234336900600402

Chan, Steve, 'Explaining War Termination: A Boolean Analysis of Causes', *Journal of Peace Research* 40/1 (2003), 49–66. doi:10.1177/0022343303040001205

Chapman, Tim, *The Congress of Vienna, Origins, Processes and Results* (London and New York: Routledge 1998).

Chetail, Vincent and Oliver Jütersonke (2015): *Peacebuilding: A Review of the Academic Literature*. Geneva: Geneva Peacebuilding Platform. http://www.gpplatform.ch.

Clayton, Anthony, *The Wars of French Decolonization* (London and New York: Routledge 2014).

Cordesman, Anthony H., George Sullivan, and William D Sullivan, *Lessons of the 2006 Israeli-Hezbollah War* (Washington, DC: CSIS 2007).

Cordesman, Anthony H. and Abraham R. Wagner, *The Gulf War* (Boulder, Col.: Westview 1996).

Croissant, Michael P., *The Armenia-Azerbaijan Conflict: Causes and Implications* (Westport, Connecticut: Greenwood Publishing Group 1998).

Dixon, Jeffrey, 'Emerging Consensus: Results from the Second Wave of Statistical Studies on Civil War Termination', *Civil Wars* 11/2 (2009), 121–36. doi:10.1080/13698240802631053

Donner, Fred M., *The Early Islamic Conquests* (Princeton: Princeton University Press 2014).

Doyle, Michael and Nicholas Sambanis, 'International Peacebuilding: A Theoretical and Quantitative Analysis', *American Political Science Review* 94/4 (2000), 779–801. doi:10.2307/2586208

Edwards, Jason, 'Foucault and the Continuation of War', in Avery Plaw ed., *The Metamorphoses of War* (Amsterdam: Rodopi 2012), 21–40

Emmers, Ralf, *Geopolitics and Maritime Territorial Disputes in East Asia* (Abingdon and New York: Routledge 2010).

Esdaile, Charles, *The Peninsular War* (London: Palgrave Macmillan 2003).

Fearon, James D., 'Why Do Some Civil Wars Last So Much Longer Than Others?', *Journal of Peace Research* 41/3 (2004), 275–301. doi:10.1177/0022343304043770

Fearon, James D. and David D. Laitin, 'Ethnicity, Insurgency, and Civil War', *American Political Science Review* 97/1 (2003), 75–90. doi:10.1017/S0003055403000534

Finlan, Alastair, *The Gulf War 1991* (Oxford: Osprey Publisher 2003).

Förster, Stig, Wolfgang J. Mommsen, and Ronald Edward Robinson, *Bismarck, Europe, and Africa: The Berlin Africa Conference 1884–1885 and the Onset of Partition* (Oxford: Oxford University Press 1989).

Foster, James L. and Garry D. Brewer, 'And the Clocks Were Striking Thirteen: The Termination of War', *Policy Sciences* 7/2 (1976), 225–43. doi:10.1007/BF00143917

Foucault, Michel, *Society Must Be Defended. Lectures at the College De France* (London: Penguin Books 2004).

Fox, William T. R., 'The Causes of Peace and Conditions of War', *The Annals of the American Academy of Political and Social Science* 392/1 (1970), 1–13. doi:10.1177/000271627039200102

Freire, Maria Raquel, *Conflict and Security in the Former Soviet Union: The Role of the OSCE* (Burlington, VT: Ashgate 2003).

Friedberg, Aaron L., *Beyond Air-to-Sea Battle. The Debate over US Military Strategy in Asia* (Abingdon: Routledge/IISS 2014).

Fuller, J. F. C., *The Conduct of War 1789–1961* (New York: Da Capo Press 1962).

Ganguly, Sumit, *The Origins of War in South Asia* (Boulder, CO.: Westview 1986).

Gates, David, *The Spanish Ulcer. A History of the Peninsular War* (New York: Da Capo Press 2001).

Gawrych, George W., *The Albatross of Decisive Victory: War and Policy between Egypt and Israel in the 1967 and 1973 Arab-Israeli Wars* (New York: Greenwood Press 2000).

Ghobarah, H., Paul Huth, and Bruce Russet, 'Civil Wars Kill and Maim People, Long after the Fighting Stops', *American Political Science Review* 97/2 (2003), 189–202. doi:10.1017/S0003055403000613

Glover, Michael, *Legacy of Glory, the Bonaparte Kingdom of Spain* (New York: Scribner's and Sons 1971).

Glover, Michael, *The Peninsular War 1807–1814: A Concise Military History* (New York and London: Penguin 2001).

Goemans, H. E., *War and Punishment. The Causes of War Termination and the First World War* (Princeton: Princeton University Press 2000).

Gulik, Edward Vose, *Europe's Classical Balance of Power* (New York and London: W. N. Norton 1955).

Györkei, Jenö, Alexandr Kirov, and Miklos Horvath, *Soviet Military Intervention in Hungary, 1956* (New York: Central European University Press 1999).

Halperin, Morton, *Limited War in the Nuclear Age* (New York: John Wiley 1963).

Halperin, Morton, 'War Termination as a Problem in Civil-Military Relations', *The Annals of the American Academy of Political and Social Science* 392/1 (1970), 86–95. doi:10.1177/000271627039200109

Hammel, Eric, *Six Days in June: How Israel Won the 1967 Arab-Israeli War* (London and New York: Simon & Schuster 1992).

Handel, Michael, 'The Study of War Termination', *Journal of Strategic Studies* 1/1 (1978), 51–75. doi:10.1080/01402397808436989

Harrison, Mark, *The Economics of World War II: Six Great Powers in International Comparison* (New York: Cambridge University Press 1998).

Heginbotham, Eric, et al., *The U.S.-China Military Scorecard. Forces, Geography and the Evolving Balance of Power 1996–2017* (Santa Monica, CA: The Rand Corporation 2015).

Hegre, Håvard, 'The Duration and Termination of Civil Wars', *Journal of Peace Research* 41/3 (2004), 243–52. doi:10.1177/0022343304043768

Hegre, Håvard, Tanja Ellingsen, Scott Gates, and Nils Petter Gleditsch, 'Toward a Democratic Civil Peace? Democracy, Political Change, and Civil War, 1816–1992', *American Political Science Review* 95 (2001), 33–48

Hobson, J. A., *Imperialism. A Study* (London: George Allen and Unwin 1905).

Holbraad, Carsten, *Superpowers and International Conflict* (London: Palgrave MacMillan 1979).

Iklé, Fred Charles, *Every War Must End* (New York: Columbia University Press 1991).

Jansen, Jan C. and Jürgen Osterhammel, *Decolonization: A Short History* (Princeton: Princeton University Press 2017).

Jasiukenaite, Berta, 'The Conception of the "New Wars". A Question of Validity', *Lithuanian Annual Strategic Review* 8/1 (2010), 25–41.

Jomini, Antoine-Henry, *The Political and Military History of the Campaign of Waterloo* (New York: Van Nostrand 1864).

Kaldor, Mary, *The New and Old Wars* (Cambridge: Polity 2006).

Kaldor, Mary, 'In Defence of New Wars', *Stability: International Journal of Security and Development* 2/1 (2013), 1–16. doi:10.5334/sta.at

Kecskemeti, Paul, *Strategic Surrender: The Politics of Victory and Defeat* (Stanford, CA.: Stanford University Press 1958).

Kennedy, Hugh, *The Great Arab Conquests: How the Spread of Islam Changed the World We Live In* (Cambridge, Mass.: Da Capo Press 2007).

Kirchberger, Sarah, *Assessing China's Naval Power. Technological Innovation, Economic Constraints, and Strategic Implications* (Berlin and Heidelberg: Springer Publ. 2015).

Kissinger, Henry A., *Nuclear Weapons and Foreign Policy* (New York: Harper & Brothers 1957).

Kissinger, Henry A., *Years of Upheaval* (Boston: Little Brown & Co. 1982).

Kissinger, Henry A., *Diplomacy* (London and New York: Simon & Schuster 1995).

Klingberg, Frank L., 'Predicting the Termination of War: Battle Casualties and Population Losses', *Journal of Conflict Resolution* 10/2 (1966), 129–71. doi:10.1177/002200276601000201

Kostanyan, Hrant and Stefan Meister, *Ukraine, Russia and the EU. Breaking the Deadlock in the Minsk Process* (Brussels: CEPS working document 423 2016).

Krasner, Stephen, 'Compromising Westphalia', *International Security* 20/3 (1995), 115–51. doi:10.2307/2539141

Krause, Joachim ed, *Iran's Nuclear Programme. Strategic Implications* (Abingdon and New York: Routledge 2012).

Krause, Joachim, 'Assessing the Danger of War: Parallels and Differences between Europe in 1914 and East Asia in 2014', *International Affairs* 90/6 (2014), 1421–51. doi:10.1111/1468-2346.12177

Krause, Joachim and Charles King Mallory IV, *International State Building and Reconstruction Efforts. Experience Gained and Lessons Learned* (Farmington Hills: Budrich Publishers 2010).

Kreutz, Joakim, 'How and When Armed Conflicts End: Introducing the UCDP Conflict Termination Dataset', *Journal of Peace Research* 47/2 (2010), 243–50. doi:10.1177/0022343309353108

Kroenig, Matthew, *A Strategy for Deterring Russian De-Escalation Strikes* (Washington, D.C.: The Atlantic Council 2018).

Langer, William L., *The Diplomacy of Imperialism: 1890–1902* (New York: A Knopf 1951).

Laqueur, Walter (July 1975): 'The Origins of Guerrilla Doctrine' *Journal of Contemporary History* 10 (3): 341–82 doi:10.1177/002200947501000301

Legier-Topp, Linda, *War Termination: Setting Conditions for Peace* (Carlisle Barracks, PA: Army War College 2009).

Licklider, Roy, 'The Consequences of Negotiated Settlements in Civil Wars, 1945–1993', *American Political Science Review* 89/3 (1995), 681–90. doi:10.2307/2082982

Mac Ginty, Roger, *International Peacebuilding and Local Resistance: Hybrid Forms of Peace* (New York and London: Palgrave MacMillan 2011).

Mason, David and Patrick Fett, 'How Civil Wars End: A Rational Choice Approach', *Journal of Conflict Resolution* 40/4 (1996), 546–68. doi:10.1177/0022002796040004002

Massoud, Tansa George, 'War Termination', *Journal of Peace Research* 33/4 (1996), 491–196. doi:10.1177/0022343396033004009

Mazaar, Michael J, *Mastering the Grey Zone. Understanding a Changing Era of Conflict* (Carlisle Barracks: US Army War College 2015).

Morgenthau, Hans J., *Politics among Nations. The Struggle for Power and Peace* (New York: A. Knopf 1963).

Münkler, Herfried, *The New Wars* (Cambridge: Polity 2005).

Nagl, John A., *Learning to Eat Soup with a Knife: Counterinsurgency Lessons from Malaya and Vietnam* (Chicago: Chicago University Press 2005).

Nicholson, Helen, *The Crusades* (Westport, CT and London: Greenwood Publishing Group 2004).

Norman, Brown, W., *The United States and India, Pakistan, Bangladesh* (Cambridge, Mass.: Harvard University Press 1972).

Oren, Michael, *Six Days of War* (Oxford: Oxford University Press 2002).

Osiander, Andreas, 'Sovereignty, International Relations, and the Westphalian Myth', *International Organization* 55/2 (2001), 251–87. doi:10.1162/00208180151140577

Pakenham, Thomas, *Scramble for Africa: The White Man's Conquest of the Dark Continent from 1876–1912* (New York: Avon Books 1992).

Paris, Roland, *At War's End. Building Peace After Civil Conflict* (Cambridge: Cambridge University Press 2004).

Paris, Roland and Timothy Sisk, *The Dilemmas of Statebuilding: Confronting the Contradictions of Postwar Peace Operations* (Abingdon: Routledge 2009).

Paul, T.V., Richard J. Harknett, and James J. Wirtz, *The Absolute Weapon Revisited: Nuclear Arms and the Emerging International Order* (Ann Arbor, MI: The University of Michigan Press 2000).

Pettersson, Therése and Peter Wallensteen, 'Armed Conflicts, 1946–2014', *Journal of Peace Research* 52/4 (2015), 536–50. doi:10.1177/0022343315595927

Pflanze, Otto ed, *The Unification of Germany, 1848–1871* (Huntington, NY: R. E. Krieger 1979).

Pillar, Paul R., *Negotiating Peace. War Termination as a Bargaining Process* (Princeton: Princeton University Press 2014).

Poggi, Gianfranco, *The Development of the Modern State. A Sociological Introduction* (Stanford, Cal.: Stanford University Press 1978).

Pollack, Kenneth, *Arabs at War: Military Effectiveness, 1948–1991* (Lincoln: University of Nebraska Press 2004).

Powell, Colin, *My American Journey. An Autobiography* (New York: Ballantine Books 1995).

Quandt, William B., *Peace Process: American Diplomacy and the Arab-Israeli Conflict since 1967* (Washington, D.C.: Brookings Institution Press 2005).

Reed, James W., *Should Deterrence Fail: War Termination and Campaign Planning* (Newport: US Naval War College 1992). http://www.dtic.mil/cgi-bin/GetTRDoc?Location=U2&doc=GetTRDoc.pdf&AD=ADA253154

Regan, Patrick, 'Conditions for Successful Third Party Interventions', *Journal of Conflict Resolution* 40/1 (1996), 336–59. doi:10.1177/0022002796040002006

Riall, Lucy, *The Italian Risorgimento: State, Society, and National Unification* (Abingdon: Routledge 1994).

Riley-Smith, Jonathan, *What Were the Crusades?* (London: Palgrave Macmillan 2009).

Rose, Gideon, *How Wars End: Why We Always Fight the Last Battle* (New York,: Simon & Schuster 2010).

Rotenberg, Gunther E., *The Art of Warfare in the Age of Napoleon* (London: B. T. Batsford 1977).

Sambanis, Nicholas, 'Partition as a Solution to Ethnic War: An Empirical Critique of the Theoretical Literature', *World Politics* 52/4 (2000), 437–83. doi:10.1017/S0043887100020074

Sambanis, Nicholas, 'Do Ethnic and Nonethnic Civil Wars Have the Same Causes? A Theoretical and Empirical Inquiry (Part 1)', *Journal of Conflict Resolution* 45/3 (2001), 259–82. doi:10.1177/0022002701045003001

Sambanis, Nicholas, 'Expanding Economic Models of Civil War Using Case Studies', *Perspectives on Politics* 2/2 (2004), 259–79. doi:10.1017/S1537592704040149

Sarkees, Meredith Reid, 'The Correlates of War Data on War: An Update to 1997', *Conflict Management and Peace Science* 18/1 (2000), 123–44. doi:10.1177/073889420001800105

Schmitt, Carl, *The Concept of the Political* (Chicago: University of Chicago Press 1996).

Schroeder, Paul W., *Metternich's Diplomacy at Its Zenith 1820–1823* (Austin: University of Texas Press 1962).

Schroeder, Robin, 'Not Too Little, but Too Late. ISAF's Strategic Restart of 2010 in Light of the Coalition's Previous Mistakes', in Joachim Krause and Charles King Mallory, IV eds., *Afghanistan, Pakistan, and Strategic Change* (Abingdon and New York: Routledge 2014), 19–69.

Showalter, Dennis E., *The Wars of German Unification* (London: Bloomsbury Academic 2004).

Shugart, Thomas, *First Strike. China's Missile Threat to U.S. Bases in Asia* (Washington, D.C.: Center for a new American Security 2017).

Slantchev, Branislav, 'The Principle of Convergence in Wartime Negotiations', *American Political Science Review* 97/4 (2003), 621–32. doi:10.1017/S0003055403000911

Smith, Simon C, *Ending Empire in the Middle East: Britain, the United States and Post-War Decolonization, 1945–1973* (New York and London: Routledge 2013).

Smith, Tony, 'The French Colonial Consensus and People's War, 1946–58', *Journal of Contemporary History* 9/4 (1974), 217–47. doi:10.1177/002200947400900410

Smith, Tony, 'A Comparative Study of French and British Decolonization', *Comparative Studies in Society and History* 20/1 (1978), 70–102. doi:10.1017/S0010417500008835

Stam, Allan C., *Win, Lose, or Draw. Domestic Politics and the Crucible of War* (Ann Arbor: University of Michigan Press 1996).

Stanley, E. and J. Sawyer, 'The Equifinality of War Termination: Multiple Paths to Ending War', *Journal of Conflict Resolution* 53/5 (2009), 651–76. doi:10.1177/0022002709343194

Stanley, Elizabeth A., 'Ending the Korean War: The Role of Domestic Coalition Shifts in Overcoming Obstacles to Peace", *International Security* 34/1 (2009a), 42–82. doi:10.1162/isec.2009.34.1.42

Stanley, Elizabeth A., *Paths to Peace: Domestic Coalition Shifts, War Termination and the Korean War* (Stanford, CA: Stanford University Press 2009b).

Stanley, Elizabeth A. and John P. Sawyer, 'The Equifinality of War Termination Multiple Paths to Ending War', *Journal of Conflict Resolution* 53/5 (2009), 651–76. doi:10.1177/0022002709343194

Stein, Janice Gross, 'Proxy Wars: How Superpowers End Them: The Diplomacy of War Termination in the Middle East', *International Journal* 35/3 (1980), 478–519.

Stevenson, David, *Cataclysm. The First World War as Political Tragedy* (New York: Basic Books 2004).

Szabo, Franz, *The Seven Years War in Europe, 1756–1763* (Abigdon and New York: Routledge 2008).

Tertrais, Bruno, 'The Demise of Ares. The End of War as We Know It', *The Washington Quarterly* 35/3 (2012), 7–22. doi:10.1080/0163660X.2012.703521

Teschke, Bruno, *The Myth of 1948. Class, Geopolitics and the Making of Modern International Relations* (London and New York: Verso 2003).

Townshend, Charles, *The Oxford History of Modern Warfare* (Oxford: Oxford University Press 2005).

Tyerman, Christopher, *The Crusades. A Brief Insight* (London and New York: Sterling 2007).

van Tol, Jan, Mark Gunzinger, Andrew Krepinevich, and Jim Thomas, *Air-Sea Battle. A Point of Departure Operational Concept* (Washington, D.C.: Center for Strategic and Budgetary Assessments 2010).

von Clausewitz, Carl, *On War* (Princeton: Princeton University Press 1989).

Wagner, R. Harrison, 'Bargaining and War', *American Journal of Political Science* 44/3 (2000), 469–84. doi:10.2307/2669259

Walter, Barbara, *Committing to Peace: The Successful Settlement of Civil Wars* (Princeton, NJ: Princeton University Press 2002).

Wawro, Geoffrey, *The Austro-Prussian War* (Cambridge: Cambridge University Press 1996).

Wawro, Geoffrey, *The Franco-Prussian War: The German Conquest of France* (Cambridge: Cambridge University Press 2005).

Weinberg, Gerhard L., *A World at Arms. A Global History of World War II* (Cambridge: Cambridge University Press 1994).

Wesseling, Henri L., *Divide and Conquer: The Partition of Africa, 1880–1914* (Westport, CT: Praeger 1996).

White, Hugh, *The China Choice. Why America Should Share Power* (Victoria, AUS: Black Inc. 2012).

Williams, E. N., *The Ancient Regime in Europe: Government and Society in the Major States 1648–1789* (London: Harper & Row 1970).

Windsor, Philip and Adam Roberts, *Czechoslovakia 1968: Reform, Repression and Resistance* (London/New York: Chatto and Windus/Columbia University Press 1969).

Wittmann, Donald, 'How Wars End. A Rational Model Approach', *Journal of Conflict Resolution* 23/4 (1979), 743–63. doi:10.1177/002200277902300408

Zhang, Xiaoming, 'China's 1979 War with Vietnam: A Reassessment', *The China Quarterly* 184/4 (2005), 851–74. doi:10.1017/S0305741005000536

In pursuit of accountability during and after war

Thomas Obel Hansen

ABSTRACT
This article aims to identify and elaborate the causes and ramifications of applying transitional justice, in particular accountability measures, to situations of war. It focuses on the correlations between peace and justice – and hence an important perspective on the question 'how do wars end'. The article seeks to understand some of the main challenges associated with pursuing accountability for crimes committed in contemporary forms of conflict, including civil wars and abuses committed by major powers in armed conflict.

Introduction

Transitional justice developed as a set of tools – and a conceptual framework – to address past abuses committed by authoritarian regimes, but is increasingly applied to situations of armed conflict, both during and after war has ended. The move from viewing transitional justice as primarily involving the judicial and quasi-judicial responses by newly installed democratic regimes to confront abuses committed by past undemocratic regimes to viewing transitional justice as relevant to a broad range of situations, including situations of past and ongoing armed conflict, raises a series of important questions. Notably, should transitional justice be viewed as a tool of conflict resolution, and if so how effective it is? Furthermore, as transitional justice, in particular accountability measures, aim at regulating the conduct of parties to armed conflict, for example by deterring the commission of crimes, what challenges do such measures face in light of the nature of contemporary types of conflict?

Although these questions are increasingly being addressed by transitional justice scholarship, much is yet to be understood. For example, whereas the interplay between justice and peace processes has gained some attention in recent studies, less attention has been paid to situations where the hostilities are not brought to an end through negotiated settlements, for example in the context of major powers' military interventions. More generally, the literature on transitional justice is yet to fully appreciate what new, and possibly unique, challenges emerge when pursuing accountability for crimes committed in contemporary types of armed conflict.

This article sets out to address some of these developments and challenges to the field of transitional justice. The article primarily aims to identify and elaborate the causes and ramifications of applying transitional justice, in particular accountability measures, to situations of war. It does so by first outlining developments in the field of transitional justice, including understandings of the correlations between peace and justice – and hence an important perspective on the question 'how do wars end'. In this regard, the article demonstrates how mainstream understandings have changed dramatically over time, from assuming that pursuing (retributive) justice poses a risk to peace, to assuming that justice is a prerequisite for building sustainable peace, and later towards an understanding that such binary claims concerning the synergies between peace and justice are too simplistic. Following this review of claims made in the literature, the article next sets out some of the main challenges associated with pursuing accountability for crimes committed in contemporary forms of conflict. This involves a discussion of the dilemmas arising when pursuing accountability for large-scale abuses committed during civil war and while conflict is ongoing. The article also addresses the challenges associated with applying the current legal frameworks as well as institutional challenges to promoting accountability for crimes committed in armed conflict. Finally, the article examines a topic overlooked in many accounts of transitional justice relating to the specific challenges associated with pursuing accountability for abuses committed by major powers in armed conflict.

Accordingly, a key objective of this article is to clarify the potential of transitional justice to address situations of war, thereby moving beyond its foundation as tool for truth and justice in democratic transitions. It argues that in so doing we need to distinguish between transitional justice's ability to contribute to ending war and to regulate the conduct of parties to armed conflict. In this sense, the article contributes to the broader debate about how transitional justice can be utilised to address the complexities of armed conflict. As such, the article observes that whereas transitional justice provides a relevant framework for advancing accountability in situations of armed conflict, transitional justice faces significant challenges addressing the particular conditions of contemporary forms of conflict which require further exploration and clarification.

A generation shift? The story of how transitional justice developed from a tool of democratisation to a tool of conflict resolution

Grounded in a merger of human rights advocacy and the 'transition to democracy' literature of the late 1980s and early 1990s, the field of transitional justice emerged in the context of the so-called third wave of democratisation.[1] The early field primarily focused on understanding how the new democracies of Latin America and East and Central Europe could use justice tools to respond to the massive human rights abuses committed under previous authoritarian or totalitarian regimes.

One key assumption in the early field was that the political transition created a window of opportunity for rendering justice for the victims of past abuses. At the same time, it was thought that transitional justice could help consolidate the new democratic order and entrench the rule of law.[2] However, early commentaries also tended to accept that the selfsame justice processes could jeopardise democratisation if they failed to operate on the conditions set by the political transition, in particular because the former elites usually maintained influence both during and after the transition.[3] Accordingly, the 'transition' – seen as a unique and confined moment in time – was assumed to present both opportunities and limitations to the kind of justice that could be rendered in these so-called paradigmatic transitions.[4] Transitional justice, it was suggested, can advance the liberal transformation, including consolidating democratic principles and the rule of law in the long term, but in so doing the process itself may need to compromise with the rule of law standards of ordinary times due to the unique circumstances in which transitional justice operates.[5] In other words, transitional justice was seen as something fundamentally distinct from justice in 'ordinary times'.[6] Often framed as a question of truth versus justice, early debates about transitional justice tended to centre around the question of whether new democracies are best served by utilising criminal justice processes or considering other responses to the past abuses, in particular truth commissions.[7]

---

[1] The term was coined by Huntington. See Samuel Huntington, *The Third Wave: Democratization in the Late Twentieth Century* (Norman OK: University of Oklahoma Press 1991). Notable studies of the transitions to democracy of this era include Guillermo O'Donnell and Philippe Schmitter, *Transitions from Authoritarian Rule: Tentative Conclusions about Uncertain Democracies* (Baltimore MD: The Johns Hopkins University Press 1986); Juan J. Linz and Alfred Stepan, *Problems of Democratic Transition and Consolidation: Southern Europe, South America, and Post-Communist Europe* (Baltimore MD: The Johns Hopkins University Press 1996).

[2] See, for example, the studies in Neil J. Kritz (ed.), *Transitional Justice: How Emerging Democracies Reckon with Former Regimes*, Volume I and II (Washington DC: United States Institute of Peace Press 1995).

[3] See, for example, Carlos Nino, 'Response: The Duty to Punish Past Abuses of Human Rights into Context: The Case of Argentina', in Kritz (ed.), *Transitional Justice*, Volume I, 417; Diane F. Orentlicher, 'A Reply to Professor Nino', in Kritz (ed.), *Transitional Justice*, Volume I, 437.

[4] On the concept of 'paradigmatic transitions', and how transitional justice emerged as a tool to promote justice in these types of transitions, see, for example, Paige Arthur, 'How "Transitions" Reshaped Human Rights: A Conceptual History of Transitional Justice', *Human Rights Quarterly* 31/2 (2009), 321–367.

[5] On the uniqueness of transitional justice and the nature and ramifications of such compromises, see further Ruti Teitel, *Transitional Justice* (Oxford: OUP 2000).

[6] One notable exception to this understanding involves Eric Posner and Adrian Vermeule, 'Transitional Justice as Ordinary Justice', *Harvard Law Review* 117 (2003), 762–825.

[7] See, for example, José Zalaquett, 'Balancing Ethical Imperatives and Political Constraints: The Dilemma of New Democracies Confronting Past Human Rights Violations', in Kritz (ed.), *Transitional Justice*, Volume I, 203–206.

Since then, transitional justice has developed enormously. One particular important development, which I have referred to elsewhere as the 'horizontal expansion of transitional justice',[8] involves the extension of transitional justice discourse to justice processes aimed at addressing abuses (and more broadly the roots of conflict) in a variety of situations not characterised by a liberalising political transition. Key among these are situations of armed conflict, frequently of internal nature but almost equally frequent with some form of regional or international dimensions.

Reflecting this change in the field's focus, transitional justice is now typically defined in ways that embrace justice after authoritarian rule as well as justice after war. According to Roht-Arriaza, for example, transitional justice can be understood as a 'set of practices, mechanisms and concerns that arise following a period of conflict, civil strife or repression, and that are aimed directly at confronting and dealing with past violations of human rights and humanitarian law'.[9] However, as discussed further below in this article, attempts at rendering justice for serious crimes increasingly occur, not only after war has ended, but while it is still ongoing, raising important questions as to how justice and peace processes interact.

From a scholarly perspective, one way of looking at the above is to say that the field has developed as a consequence of the fact that the type of legal and quasi-legal measures referred to as transitional justice when occurring in paradigmatic transitions are now increasingly utilised in situations such as Sierra Leone, Colombia and Uganda where the (main) transition in question is not one from authoritarianism to democracy but one from some form of armed conflict to (relative) peace and stability. In a sense, the fact that justice processes in these and other countries ostensibly undergoing peaceful transformation are now being debated as transitional justice could be seen as a generation shift, reflecting a change in world affairs with fewer democratic transitions and where large-scale abuses increasingly take place in the context of armed conflict, in particular civil wars and other forms of internal strife.[10]

The move towards conceptually embracing transitions that predominantly concern an already existing or attempted move from armed conflict to (relative) peace brings into question whether one can operate with one common theory of what transitional justice is and what it can achieve, or if

---

[8]Thomas Obel Hansen, 'The Vertical and Horizontal Expansion of Transitional Justice: Explanations and Implications for a Contested Field', in Susanne Buckley-Zistel et al. (eds.), *Transitional Justice Theories* (London: Routledge 2013), 105–124.

[9]Naomi Roht-Arriaza, 'The New Landscape of Transitional Justice', in Naomi Roht-Arriaza and Javier Mariezcurrena (eds.), *Transitional Justice in the Twenty-First Century: Beyond Truth versus Justice* (Cambridge: CUP 2006), 1–16, at 2.

[10]See further Andrew Reiter et al., 'Transitional Justice and Civil War: Exploring New Pathways, Challenging Old Guideposts', *Transitional Justice Review* 1/1 (2012), 137–169 (noting at 138 that whereas the number of post-authoritarian transitions is waning, 'civil wars continue to proliferate around the world, offering new opportunities for transitional justice').

there is a need for developing more context-specific frameworks, and if so how.[11] It also begs the question whether all efforts to promote accountability for serious crimes should be understood as 'transitional justice'.[12] One particularly important question concerns the goals that transitional justice should advance. On the one hand, some scholars imply that a normative framework which emphasises democratic consolidation as a key outcome of transitional justice may also be suitable for addressing other types of transition currently addressed by field.[13] On the other hand, some commentators suggest that the expansion of the types of situations addressed by transitional justice must be matched by an expansion of the goals of transitional justice.[14] As discussed in further detail below, the potential of transitional justice to advance peace-building is now pointed to as a central goal.

The correlations between peace and justice

The first-generation argument: peace versus justice

From the outset, transitional justice was particularly occupied with a question key to understanding 'how wars end', namely a perceived tension between peace and justice, specifically whether pursuing criminal justice for past crimes would endanger peace and stability. For example, in an early account of transitional justice, Zalaquett concluded that actors such as the armed forces, who may be opposed to transitional justice, will 'determine the scope of governmental action inasmuch as they may affect its stability or force its hand, depending on the actual human rights policy the government attempts to carry out'.[15] Zalaquett viewed the tension between peace and justice as a tension between principles and pragmatic concerns: 'Ethical principles provide guidance but no definite answer. Political leaders cannot afford to be moved only by their convictions, oblivious to real-life constraints, lest in the end the very ethical principles they wish to uphold suffer

---

[11]See further Thomas Obel Hansen, 'Transitional Justice: Toward a Differentiated Theory', *Oregon Review of International Law* 13/1 (2011), 1–46.

[12]Some scholars have warned against confusing transitional justice with international criminal law. See, for example, Jens Iverson, 'Transitional Justice, Jus Post Bellum and International Criminal Law: Differentiating the Usages, History and Dynamics', *International Journal of Transitional Justice* 7/3 (2013), 413–433.

[13]See, for example, Fionnuala Ni Aoláin and Colm Campbell, 'The Paradox of Transition in Conflicted Democracies', *Human Rights Quarterly* 27 (2005), 172–213 (noting at 174 that the 'end goal of transition in conflicted democracies is the same as that in paradigmatic transitions, namely the achievement of a stable (and therefore peaceful) democracy').

[14]See, for example, Phil Clark, 'Establishing a Conceptual Framework: Six Key Transitional Justice Themes', in Phil Clark and Zachary Kaufman (eds.), *After Genocide: Transitional Justice, Post-Conflict Reconstruction and Reconciliation in Rwanda and Beyond* (London: Hurst 2008), 191–205 (arguing that transitional justice should ideally aim to achieve all of the following: reconciliation, peace, justice, healing, forgiveness and truth).

[15]José Zalaquett, 'Confronting Human Rights Violations Committed by Former Governments: Principles Applicable and Political Constraints', in Kritz (ed.), *Transitional Justice*, Volume I, 3–31, at 17.

because of a political or military backlash'.[16] Even scholars such as Nino who strongly called for criminal justice measures in countries such as Argentina undergoing transition in the late 1980s and early 1990s acknowledged that limiting the scope of criminal justice could be necessary because the armed forces 'still retained a monopoly on state coercion and were united in their opposition to the trials', and, therefore, had the capacity to threaten the 'only force backing the trial – the democratic system.'[17]

Accordingly, a key concern in the early writings on transitional justice was that efforts to prosecute those responsible for past abuses could pose an existential threat to the new democracy because the transition, often facilitated through elite pacts, usually did not completely rid the former rulers responsible for the abuses of power. In short, the mainstream view was that criminal justice should ideally be pursued to address the crimes of the past, but that the particular – or even unique – circumstances of the political transition create significant limitations to how much justice can be achieved in practice.

The second-generation argument: no peace without justice

A radical development occurred in the 2000s as numerous scholars and practitioners alike started suggesting that the pursuit of justice – including criminal justice – should be viewed, not as an obstacle, but as a *prerequisite* for peace.[18] Rather than posing a potential threat to peace and stability as had been assumed in much of the earlier writings on transitional justice, the mainstream view became that only by pursuing justice for past crimes can sustainable peace be achieved. Without justice for the abuses of the past, it was suggested, peace would be only a temporary condition since the grievances of the past would re-surface and jeopardise any settlement made in the transition.[19] 'No peace without justice' became the dominant mantra of the time.

---

[16]Zalaquett, 'Balancing Ethical Imperatives and Political Constraints', 203–206, at 205.

[17]Carlos Nino, 'Response: The Duty to Punish Past Abuses of Human Rights into Context: The Case of Argentina', 417–436, at 421.

[18]Of course, the change in perceptions did not occur overnight and there were dissenting voices. However, a highly promoted and widely attended 2007 conference in Nuremberg, entitled 'Building a Future on Peace and Justice' appears as a key turning points whereby the view that peace and justice are not only compatible notions but are mutually dependent outcomes of a transition was consolidated as the mainstream.

[19]For example, writing in 2002 – the same year that the International Criminal Court became operational – M. Cherif Bassiouni, a leading international lawyer, argued that 'if peace is not intended to be a brief interlude between conflicts, it must, in order to avoid future conflict, encompass what justice is intended to accomplish: prevention, deterrence, rehabilitation, and reconciliation'. See Bassiouni, 'Accountability for Violations of International Humanitarian Law and Other Serious Violations of Human Rights', in Bassiouni (ed.), *Post-Conflict Justice* (New York: Transnational Publishers 2002), 3–54, at 9.

The proponents of 'peace and justice' often suggested that the 'peace versus justice' argument was based on a false dichotomy. For example, writing in 2006, Ellis argued: 'decisions not to prosecute are often premised on a misguided belief that it is necessary to choose between justice and peace, but this is a false choice. There can be no lasting peace without justice, and justice cannot exist without accountability. Peace cannot exist unless society first deals with the deep divisions created by human rights abuses'.[20] Impunity, rather accountability, thus came to be seen as the threat to peace – and it continues to be so by many advocates and scholars of transitional justice.

The understanding that peace and justice are mutually dependent and reinforcing moved transitional justice form the periphery of the democratisation literature to the very centre of debates about peace and human rights, in that way presenting new expectations to how wars should end. In a sense, as Nagy observes, it made transitional justice a 'global project', endorsed by almost everyone because it was seen to be 'good' in almost all ways.[21] However, despite broad consensus that long-term peace can only be achieved if justice for past abuses is done, surprisingly few attempts were made in this period to justify these claims with reference to empirically grounded research. In other words, the proponents of 'peace and justice' made lofty assertions, often advocating for far-reaching criminal justice processes, to address serious crimes committed by past authoritarian regime or, increasingly, in the context of armed conflict. But they often failed to demonstrate why the correlations between peace and justice should be perceived so straightforward. In particular, many proponents of 'peace and justice' seemed to overlook the importance of understanding how political dynamics, both local and international, impact justice and peace processes and that neither peace nor justice processes are static, but rather dynamic and multifaceted processes that may interact in multiple, complex ways and may differ significantly over time and space.

The third-generation argument: towards understanding peace and justice complexities?

One important trend in more contemporary studies of transitional justice involves the turn to empirical research on peace and justice correlations, and more broadly the development of more sophisticated methodologies to

---

[20]Mark Ellis, 'Combating Impunity and Enforcing Accountability as a Way to Promote Peace and Stability – The Role of International War Crimes Tribunals', *Journal of National Security and Policy* 2/1 (2006), 111–164.
[21]See Rosemary Nagy, 'Transitional Justice as Global Project: Critical Reflections', *Third World Quarterly* 29/2 (2008), 275–289, at 276.

understand the impact of transitional justice.[22] Importantly, from the late
2000s onwards, scholars increasingly started using social science methods to
undertake cross-country comparative analyses of transitional justice, developed databases on transitional justice tools, and used detailed case studies
to test the assumptions made about peace and justice correlations.[23]

However, the results of such studies often contradict each other, making
it difficult to reach clear conclusions concerning how transitional justice
impacts conflict prevention and resolution – and hence the 'end of war'.
By way of example, claiming to offer the first systematic empirical assessment of the deterrent effects of the International Criminal Court (ICC), Jo
and Simmons observe that the Court can deter perpetrators and reduce
intentional violence against civilians in civil wars, in that way contributing to
ending wars.[24] Yet, some question Jo and Simmons' methodology as well as
their conclusions concerning the ICC's potential to deter potential perpetrators, and more generally its ability to prevent atrocities.[25] Despite the turn
to empirically based research, transitional justice's contribution to ending
wars thus remains disputed. If anything, it has become clear that context
matters. Emphasising that 'there is no strict formula for timing and sequencing of peacebuilding or transitional justice activities' because both sets of
activities are dynamic and context-specific, Sriram et al. highlight ways in
which transitional justice mechanisms can both complement and contradict
various aspects of peacebuilding such as disarmament, demobilisation and
reintegration; security sector reform; and rule of law promotion.[26]

At the same time, there is increased awareness that pursuing synergies
between transitional justice and 'liberal peacebuilding' can be problematic.
Noting that a 'forceful criticism of liberal peacebuilding has developed in
recent years' which challenges its 'twin emphases on democratisation and

---

[22]To my knowledge, the first comprehensive study which elaborates methodological issues concerning
impact studies in transitional justice is the 2009 volume, *Assessing the Impact of Transitional Justice*
(edited by Hugo van der Merwe, Victoria Baxter and Audrey Chapman) (Washington DC: United
States Institute of Peace Studies 2009).

[23]See, for example, David Backer, 'Cross-National Comparative Analysis', in Hugo van der Merwe,
Victoria Baxter and Audrey Chapman (eds.), *Assessing the Impact of Transitional Justice*, 23–90.
Louise Mallinder has developed the extensive Amnesty Law Database involving information on
506 amnesty processes in 130 countries introduced since the Second World War. See further
Louise Mallinder, *Amnesty, Human Rights and Political Transitions: Bridging the Peace and Justice
Divide* (Oxford: Hart Publishing 2008).

[24]See Hyeran Jo and Beth A. Simmons, 'Can the International Criminal Court Deter Atrocity?',
International Organization 70/3 (2016), 443–475.

[25]See Jack Snyder and Leslie Vinjamuri, 'To Prevent Atrocities, Count on Politics First, Law Later',
openDemocracy, 12 May 2015, available at www.opendemocracy.net/openglobalrights/jack-
snyder-leslie-vinjamuri/to-prevent-atrocities-count-on-politics-first-law-late. For a further debate
about the ICC's claimed deterrent effect, see Jack Snyder and Leslie Vinjamuri, 'Trials and Errors:
Principle and Pragmatism in Strategies of International Justice', *International Security* 28 (2003), 5–44.

[26]Chandra Lekha Sriram, Johanna Herman and Olga Martin-Ortega, 'Beyond Justice versus Peace:
Transitional Justice and Peacebuilding Strategies', in Karin Aggestam and Annika Björkdahl (eds.),
Rethinking Peacebuilding: The Quest for Just Peace in the Middle East and the Western Balkans (New
York: Routledge 2012), 1–23, at 4.

marketisation and the presumption that democratisation and market liberal-
isation are themselves sources of peace', Sriram argues that transitional
justice could be subject to much of the same criticism, as it shares with
liberal peacebuilding a number of under-examined assumptions and unin-
tended consequences.[27] Since then, some of these assumptions and consequences have been subject to more scrutiny. Notably, Sharp argues that
'the dominant liberal transitional justice paradigm has often resulted in a
relatively narrow or thin approach to questions of justice in transition that
foregrounds physical violence, including violations of physical integrity and
civil and political rights issues more generally, while pushing questions of
economic violence and economic justice to the margins'.[28] Sharp concludes
that there are 'strong reasons to suspect that more integrated approaches to
peacebuilding and transitional justice will have the tendency to exacerbate
some of the tendencies that have given rise to these parallel critiques rather
than alleviate them'.[29]

Furthermore, it has gradually become clear that, rather than having one
singular impact on peace processes, it is necessary to discern how different
aspects of accountability processes relate to different aspects of conflict and
conflict resolution. Focusing specifically on the ICC's impact on conflict and
peace processes, Kersten observes that the type of actors targeted by the
Court in a given situation is key determinant.[30] This author's research in
Kenya similarly suggest that the Court's intervention had a significant and
complex impact on domestic politics, including the creation of new alliances
as a consequence of whom the Prosecutor decided to prosecute.[31]

Moreover, the *timing* of justice processes has emerged as a key topic, as it
has become clear that the feasibility of pursuing criminal justice in particular
can vary significantly over time. Noting that the 'Dirty Wars' in Latin America
of the 1970s and 1980s are now increasingly being addressed by local courts
after years of de facto or de jure impunity, Collins for example argues that

---

[27]Chandra Lekha Sriram, 'Justice as Peace? Liberal Peacebuilding and Strategies of Transitional Justice',
Global Society, 21/4 (2007), 579–591.

[28]Sharp further notes that liberal peacebuilding 'has at times resulted in a "top-down" approach to
justice, concerned more with the bargains between elite groups needed to sustain the political
transition than a more participatory approach to building democracy from the grassroots', and has
'tended to privilege the state, the international, and the universal, over the local, the traditional, and
the particular'. See Dustin Sharp, 'Interrogating the Peripheries;
The Preoccupations of Fourth Generation Transitional Justice', *Harvard Human Rights Journal* 26
(2013), 149–178.

[29]Dustin Sharp, 'Beyond the Post-Conflict Checklist: Linking Peacebuilding and Transitional Justice
through the Lens of Critique', *Chicago Journal of International Law* 14 (Summer 2013), 165–196.

[30]Kersten points to a range of other interactions between peace and justice, some of which are
discussed further below in this article. See further Mark Kersten, *Justice in Conflict: The Effects of the
International Criminal Court's Interventions on Ending Wars and Building Peace* (Oxford: OUP 2016), at
193–200.

[31]See Thomas Obel Hansen, 'Transitional Justice in Kenya? An Assessment of the Accountability Process
in Light of Domestic Politics and Security Concerns', *California Western International Law Journal*, 42/1
(2011), 1–35.

new opportunities for pursuing criminal justice often arise long after a democratic order has been formally established, due to institutional reforms; the impact of other transitional justice measures, especially truth-seeking; the simple passage of time; and other factors.[32]

Accordingly, there is now recognition that the contribution of justice processes to efforts to prevent or end war are more complex than those advocated in earlier phases of transitional justice scholarship. In particular, it has become clear that justice processes impact differently on peace processes in different situations; that justice processes contain a range of dynamics which may impact peace and conflict processes in multiple, sometimes contradicting ways; and that transitional justice must be viewed not as a one-off event, but rather as a range of processes that can take different directions over time and space.[33]

General challenges pursuing accountability for crimes in war

The nature of contemporary armed conflict

There is general agreement that we have witnessed a 'discernible shift from the industrialised total warfare of the first half of the twentieth century, to contemporary forms of low-intensity conflict in conditions of state breakdown'.[34] Such low-intensity conflicts – frequently intra-state, but often with regional or international dimensions – are typically characterised by less clear distinctions between combatants and civilians. Partly because of this, as Engstrom notes, whilst 'most war fatalities in the early twentieth century were military personnel, at the turn of the century most war fatalities were civilian'.[35] Civilians are often deliberately targeted as part of the broader strategies of the warring parties, including large-scale sexual violence, forcible recruitment and displacement and other international crimes. Besides the necessity of *resolving* the conflicts that surround these abuses, there is a need for *regulating* the conduct of parties to armed conflicts through effective accountability regimes and otherwise. However, the nature of contemporary forms of conflict, in particular civil wars, presents a range of challenges pursuing accountability for the abuses frequently associated with them.

---

[32]See Cath Collins, 'The End of Impunity? "Late Justice" and Post-Transitional Justice in Latin America', in Nicola Frances Palmer et al. (eds.), *Critical Perspectives in Transitional Justice* (Cambridge: Intersentia 2012), 399–423, at 399–419.

[33]For a further discussion of the 'time' and 'space' of transitional justice, see Thomas Obel Hansen, 'The Time and Space of Transitional Justice', in Dov Jacobs et al. (eds.), *Research Handbook on Transitional Justice* (Cheltenham: Edward Elgar Publishing 2017), 34–51.

[34]Par Engstrom, 'Transitional Justice and Ongoing Conflict', in Chandra Lekha Sriram et al. (eds.), *Transitional Justice and Peacebuilding on the Ground: Victims and Excombatants* (London: Routledge, 2012), 41–61.

[35]Ibid, at 3.

Notably, it can be difficult to draw a clear distinction between victims and perpetrators in contemporary forms of conflict. This presents a number of dilemmas for transitional justice, especially accountability processes. For example, in the only trial commenced to date before the ICC relating to the conflict in Northern Uganda, Dominic Ongwen, in his capacity as a commander of the Lord's Resistance Army (LRA), stands accused of multiple war crimes and crimes against humanity, including conscription and use of child soldiers – a crime he was himself subjected to in a young age when he was abducted into the LRA while walking to school.[36] Ongwen is thus both a victim and a perpetrator, illustrating, as Drumbl notes how the 'lines between victims and victimizers in atrocity often are porous'.[37] The legitimacy of prosecuting Ongwen before the ICC has left both academics experts and local communities divided,[38] but the challenges are not unique to this case.

Another important point is that civil war contexts tend to be characterised by a higher magnitude of violent acts compared with the abuses committed under authoritarian rule which informed the early field of transitional justice.[39] This has serious ramifications for the feasibility of transitional justice. For example, in situations where significant proportions of the population have been victimised, it is unrealistic to expect that transitional justice tools such as reparations programmes can attend to the individual needs and rights of all victims. In addition, the sheer number of perpetrators may, combined with the effects of war such as the total or partial collapse of the legal system, make it impossible to pursue individual accountability in domestic courts in the post-war context for any significant proportion of perpetrators, even in situations where there is political will to do so.[40]

Despite the emergence of a system of international justice, the ability of institutions such as the ICC to prosecute perpetrators of international crimes also remains limited, in part due to lack of resources, the complexity of investigating and prosecuting international crimes and the need for State

---

[36]See the info available at https://www.icc-cpi.int/uganda/ongwen.

[37]Mark Drumbl, 'The Ongwen Trial at the ICC: Tough Questions on Child Soldiers', openDemocracy, 14 April 2015, available at https://www.opendemocracy.net/openglobalrights/mark-drumbl/ongwen-trial-at-icc-tough-questions-on-child-soldiers.

[38]For a symposium of academic commentaries on the topic, see 'The Dominic Ongwen Trial and the Prosecution of Child Soldiers – A JiC Symposium', Justice in Conflict, available at https://justiceinconflict.org/2016/04/11/the-dominic-ongwen-trial-and-the-prosecution-of-child-soldiers-a-jic-symposium/. For a discussion of the reactions to the trial in Northern Uganda, see, for example, Lino Owor Ogora, 'Just or Unjust: Mixed Reactions on whether Ongwen should be on Trial', JusticeHub, 25 April 2017, available at https://justicehub.org/article/just-or-unjust-mixed-reactions-whether-ongwen-should-be-trial.

[39]See similarly Reiter et al., 'Transitional Justice and Civil War: Exploring New Pathways, Challenging Old Guideposts', 137–169, at 139.

[40]Accordingly, tensions may arise between promoting accountability norms and respecting due process rights of suspected perpetrators, as was so clearly the case in Rwanda following the genocide and the civil war that surrounded it. See, for example, William A, Schabas, 'The Rwanda Case: Sometimes It's Impossible', in Bassiouni (ed.), Post Conflict Justice, 499–522.

cooperation and enforcement. In the DRC (Democratic Republic of the Congo) – the situation in which the ICC has indicated the largest number of individuals – seven arrest warrants have been issued to date, mostly for mid-level rebel leaders.[41] Combined with the absence of a comprehensive domestic accountability process, the vast majority of perpetrators of international crimes, especially government-connected, are thus treated to impunity.[42] Similarly, despite attempts at pursuing accountability in other situations of civil war, such as Uganda, Central African Republic, Sudan, and Libya, these only a tiny fraction of those responsible for the massive crimes committed are brought to account, and in many situations suspected perpetrators continue to wield political or military power. Yet, such attempts at rendering justice, even if limited in scope, may in some ways positively impact the conduct of parties to the hostilities. For example, some suggest that awareness of the unlawfulness of conscribing child soldiers has increased following the indictment and later conviction of Lubanga for such crimes.[43] More broadly, the existence of accountability processes can empower actors working to constrain abuses in armed conflict.

One-sided justice

War crimes and other abuses committed in war are often perpetrated by combatants on all sides to the conflict (though of course the scope and seriousness of crimes committed by different actors can vary). This presents, at least partially, a difference from the contexts of democratic transitions that informed the early field of transitional justice.[44] Complicity on both sides raises important questions as to how to ensure even-handed justice, and the consequences of failing to do so. In situations where war ends with a clear victory to one side, accountability measures at the domestic level tend to apply to the losing side only. In Rwanda, for example, the post-genocide Gacaca courts have prosecuted hundreds of thousands of genocide perpetrators, whereas members of the Rwandan Patriotic Front (RPF) responsible for war crimes and possibly crimes

---

[41]See the info available at https://www.icc-cpi.int/drc.

[42]For a recent account of the challenges associated with pursuing accountability, internationally and nationally, for crimes in the DRC, see Patryk Labuda, 'Taking Complementarity Seriously: Why is the International Criminal Court Not Investigating Government Crimes in Congo?', *Opinio Juris*, 28 April 2017, available at http://opiniojuris.org/2017/04/28/33093/?utm_source=feedburner&utm_medium=email&utm_campaign=Feed%3A+opiniojurisfeed+%28Opinio+Juris%29.

[43]For a thoughtful debate about the ICC's ability to deter specific crimes such as recruitment of child soldiers and the impact of the Lubanga ruling, see Abigail Reynolds, 'Deterring the Use of Child Soldiers in Africa: Addressing the Gap Between the Mandate of the International Criminal Court and Social Norms and Local Understandings', *Master Thesis submitted to Leiden University*, June 2016, available at https://openaccess.leidenuniv.nl/bitstream/handle/1887/41163/Reynolds,%20Abigail-s1754807-MA%20Thesis%20PS-2016.pdf?sequence = 1.

[44]As noted by Reiter et al., while 'authoritarian regime transitions tend to involve abuses by one set of actors, war tends to involve complicity on both sides'. See Reiter et al., 'Transitional Justice and Civil War: Exploring New Pathways, Challenging Old Guideposts', 139.

against humanity in the civil war that surrounded the genocide have not been brought to account.[45] Yet, problems with one-sided justice are not limited to accountability mechanisms at the national level but extend to international tribunals (and have done so since the Nuremberg trials).[46] Although the establishment of an accountability system at the international level was intended to promote accountability for State crimes,[47] the fact remains that international justice is typically only successful prosecuting those in opposition to the incumbent, including rebel forces and ousted State leaders.

Attempts by the ICC to render justice for all sides to a conflict responsible for abuses have so far proven largely unsuccessful, for a large part due to the Court's dependence on States' cooperation.[48] For example, in the situation relating to Darfur, Sudan, the Court has issued arrest warrants and summonses to appear for both government officials, including sitting head of State Omar al-Bashir, and rebel leaders, but only the latter have appeared before the Court due to the non-cooperation of the Sudanese government (and other States).[49] In the Kenyan situation, the Prosecutor deliberately pursued a strategy of targeting both sides to the 2007–08 post-election violence.[50] This strategy arguably aimed at countering the criticism arising from earlier situations such as Uganda where the Office had pursued only one party to the conflict (i.e., the LRA), notwithstanding credible allegations that the Ugandan army is also responsible for serious and large-scale violations of international law.[51] However, the attempt at rendering justice to both sides of the violence in Kenya ultimately proved unsuccessful, in part because those targeted continued to wield – or subsequently accessed –

---

[45]See, for example, René Lemarchand, 'The Politics of Memory in Post-Genocide Rwanda', in Phil Clark and Zachary Kaufman (eds.), *After Genocide: Transitional Justice, Post-Conflict Reconstruction and Reconciliation in Rwanda and Beyond* (London: Hurst Publishers 2008), 21–30.

[46]See, for example, Victor Peskin, 'Beyond Victor's Justice? The Challenge of Prosecuting the Winners at the International Criminal Tribunals for the Former Yugoslavia and Rwanda', *Journal of Human Rights* 4/2 (2005), 213–231.

[47]See further William A. Schabas, 'State Policy as an Element of International Crimes', *Journal of Criminal Law and Criminology* 98 (2008), 953–982.

[48]On the importance of State cooperation (and the ICC's limited ability to promote it), see further Rita Mutyaba, 'An Analysis of the Cooperation Regime of the International Criminal Court and its Effectiveness in the Court's Objective in Securing Suspects in its Ongoing Investigations and Prosecutions', *International Criminal Law Review* 12 (2012), 937–962; Human Rights Law Centre, University of Nottingham, 'Cooperation and the International Criminal Court Report', Report on Expert Workshop, 18–19 September 2014.

[49]See, for example, Gwen P. Barnes, 'The International Criminal Court's Ineffective Enforcement Mechanisms: The Indictment of President Omar Al Bashir', *Fordham International Law Journal* 34/6 (2011), 1585–1619. Most recently South Africa failed to arrest al-Bashir as he attended a meeting there in June 2016, leading to significant controversy, which arguably was a major reason why South Africa later announced its intention to withdraw from the ICC. See, for example, International Commission of Jurists, 'South Africa appears before ICC for failure to arrest Sudanese President Bashir – The ICJ observes the hearing', 6 April 2017, available at https://www.icj.org/south-africa-appears-before-icc-for-failure-to-arrest-sudanese-president-bashir-the-icj-observes-the-trial/.

[50]The PEV did not amount to an 'armed conflict' in humanitarian law terms, but the dynamics discussed here are nonetheless of interest.

[51]See further Sarah Nouwen and Wouter Werner, 'Doing Justice to the Political: The International Criminal Court in Uganda and Sudan', *The European Journal of International Law* 21/4 (2011), 941–965.

power and effectively mobilised against the ICC and because the Court had few remedies in the face of witness interference and lack of cooperation by the Kenyan government that followed the attempt to hold to account those in power. Due to these and other factors, all of the ICC cases relating to the post-election violence ultimately collapsed.[52]

A key challenge promoting even-handed justice is, therefore, that the effective operation of the system of international justice depends on State cooperation, but States tend to cooperate only when it is their opponents that are being targeted.[53] As Kersten notes, one key ramification of prosecuting only one party to a conflict is that international accountability systems risk creating an 'asymmetrical understanding of the causes and drivers of violence as well as the responsibility for atrocities'.[54] Yet, even if rarely even-handed and lucidly influenced by politics, the contemporary system for pursuing accountability for violations of the law on war presents progress from a rule of law perspective compared to the total impunity that characterised previous periods. The legitimacy of institutions such as the ICC cannot be simply dismissed because it largely operates on the premises of a world order that continues to privilege State sovereignty, rather than transgressing it.[55]

Blurred lines between war and peace

The starting point for bringing into play justice measures for serious crimes has historically been *after* war and/or the end of repressive regimes. However, justice tools, in particular accountability measures, are increasingly applied in 'institutionally and politically very fragile and unstable situations, before any discernible transition' from war to peace.[56] The fact that

---

[52]See further Thomas Obel Hansen, 'The International Criminal Court in Kenya: Three Defining Features of a Contested Accountability Process and Their Implications for the Future of International Justice', *Australian Journal of Human Rights* 18/2 (2012), 187–217; Yvonne M. Dutton, 'Enforcing the Rome Statute: Evidence of (Non) Compliance from Kenya', *Indiana International and Comparative Law Review* 26/7 (2016), 7–32.

[53]However, the Prosecutor has also been faulted for not developing sufficiently clear strategies on the basis of the goals of international justice for making selection decisions. See further Margaret M. deGuzman, 'Choosing to Prosecute: Expressive Selection at the International Criminal Court', *Michigan Journal of International Law* 33/2 (2012), 265–320.

[54]Kersten, however, also notes that the ICC does not necessarily create, but rather tends to reinforce already existing narratives of 'good versus evil'. See Mark Kersten, *Justice in Conflict*, at 194–195.

[55]For an account of the legitimacy challenges facing the current system of international justice, see further Thomas Obel Hansen, 'The International Criminal Court and the Legitimacy of Exercise', in Per Andersen et al. (eds.), *Law and Legitimacy* (Copenhagen: DJOEF 2015), 73–100.

[56]Par Engstrom, 'Transitional Justice and Ongoing Conflict' (further noting that 'there has been a discernible shift from the pursuit of accountability strategies after the cessation of armed hostilities on the one hand, and in the aftermath of political transitions on the other, to attempts to achieve accountability for atrocities even before a political settlement of armed conflict has been reached', at 42.

accountability processes are now frequently brought into play in situations where war is still ongoing presents, as Engstrom notes, 'a dramatic shift in the global accountability regime'.[57] Although the practice of international judicial interventions in ongoing conflicts can be traced back to the International Criminal Tribunal for the former Yugoslavia (ICTY), this trend has become more outspoken due to the operations of the ICC.[58] Indeed, the majority of ICC investigations were launched at a time where some form of armed conflict was still ongoing in the relevant countries.

As the ICC intervenes in ongoing conflicts in places such as Uganda, Sudan, Libya, the DRC, the Central African Republic, and Mali, the question emerges as to why this change concerning the timing of accountability processes has occurred. It has been argued that the pursuit of accountability during ongoing conflict is underpinned and driven by two main underlying trends, namely the intractability of contemporary armed conflict and the dramatic expansion of the international legal architecture.[59] More generally, scholars such as Sikkink point to the existence of a 'justice cascade', understood to comprise a shift in the legitimacy of the norm of individual criminal accountability for human rights violations and an increase in criminal prosecutions – domestically and internationally – of transgressions of that norm, which has emerged for a large part due to the creativity and increased global reach of activists and human rights lawyers.[60]

However, despite the apparent normative appeal of accountability norms and the expansion of international legal regimes, there are significant challenges for international law to adequately address accountability issues in contemporary conflicts. For example, various forms of low-intensity conflict may not necessarily amount to an 'armed conflict' within the meaning of humanitarian law, which can present obstacles for pursuing individual criminal accountability for crimes committed during such conflicts. More generally, it is disputed to what extent humanitarian law applies beyond classical situations of war, including in situations labelled post-war but characterised by high levels of

---

[57]Ibid. On the topic of justice in conflict, see further Thomas Unger and Marieke Wierda, 'Pursuing Justice in Ongoing Conflict: A Discussion of Current Practice', in Kai Ambos, Judith Large et al. (eds.) *Building a Future on Peace and Justice* (Berlin: Springer 2009), 263–302.

[58]See similarly Par Engstrom, 'Transitional Justice and Ongoing Conflict' (noting that: 'the ICTY issued only few indictments during the armed conflict itself and cases only came to trial after the end of the war. Instead, the first significant attempt to pursue justice during ongoing conflict came with the indictment of Slobodan Milosevic during the NATO bombing of Kosovo', at 42.

[59]Ibid, at 3. See also Luis Moreno-Ocampo, 'Transitional Justice in Ongoing Conflicts', *International Journal of Transitional Justice* 1/1 (2007), 8–9, at 8.

[60]Kathryn Sikkink, *The Justice Cascade: How Human Rights Prosecutions Are Changing World Politics* (New York and London: W.W. Norton and Company 2011).

violence.[61] As the lines between 'war' and 'peace' have become increasingly blurred, ambiguity concerning the applicable legal frameworks also increases. As Stahn argues, the 'classical dichotomy of peace and war has lost part of its significance due to the shrinking number of inter-state wars after 1945 and the increasing preoccupation of international law with civil strife and internal armed violence'.[62] Partly due to the current interweaving of the concepts of intervention, armed conflict and peace-making, Stahn suggests that the classical rules of *jus ad bellum* and *jus in bello* should be complemented with a third branch of international law, namely rules and principles governing peace-making after conflict, referred to as *jus post bellum*.[63]

As the law currently reads, there are multiple, partly overlapping, legal frameworks regulating the conduct in armed conflict and the accountability regimes relating to it. Importantly, whereas abuses in armed conflict may, depending on the circumstances, amount to violations of international humanitarian as well human rights law, the consequences of applying the respective legal frameworks vary significantly as the former primarily triggers individual criminal accountability, whereas the latter primarily concerns State liability. Despite the emergence of notions of 'victim-centred' justice in international criminal law, victims of serious abuses committed in times of armed conflict may often be better served by pursuing accountability within a human rights framework, to the extent it applies to the situation where the abuses were committed.[64]

---

[61] As Stahn notes: 'the norms of international humanitarian law, by definition, apply only to a limited extent to the period following the cessation of hostilities. Additional Protocol I provides that the application of the Geneva Conventions and the Protocol will cease "on the general close of military operations". This moment is usually deemed to occur "when the last shot has been fired". Only selected provisions apply after the "cessation of active hostilities". A "post-conflict" duty, namely the obligation to repatriate, is activated in a classical "wartime" situation, namely before the close of military operations, which marks the date of the termination of "armed conflict". Moreover, parts of the "law of war", namely specific duties of the occupant under the laws of occupation, continue to apply in a "peacetime" situation, namely after the close of military operations. The norms of international humanitarian law are, therefore, only to a limited extent relevant to the broader process of building peace after conflict'. See Carsten Stahn, 'Jus ad bellum', 'jus in bello' ... 'jus post bellum'? – Rethinking the Conception of the Law of Armed Force, *European Journal of International Law* 17/5 (2007), 921–943, at 927.

[62] Ibid, at 923.

[63] Ibid, at 929.

[64] From a victims' perspective, pursuing accountability under human rights law may often prove the more feasible route, partly due to the relative accessibility of courts such as the ECtHR and other regional human rights courts and due to the ability of victims to bring human rights cases before domestic courts. Even if the ICC is mandated to order reparations to victims of international crimes and victims can participate in the proceedings, these regimes face significant challenges. See further Luke Moffett, 'Elaborating Justice for Victims at the International Criminal Court', *Journal of International Criminal Justice* 13 (2015), 281–311; Christine van den Wyngaert, 'Victims before International Criminal Courts: Some Views and Concerns of an ICC Trial Judge', *Case Western Reserve Journal of International Law* 44 (2011), 475–496. Human rights law, however, only applies to situations of armed conflict to the extent the armed forces have 'effective control', as will be the case for example in detention facilities. See, for example, Daragh Murray, *Practitioners' Guide to Human Rights Law in Armed Conflict* (Oxford: OUP 2016).

Beyond the legal uncertainties, the blurred lines between war and peace and the tendency to pursue justice while conflict is still ongoing raise questions concerning the reach of transitional justice, especially accountability measures, and the ramifications thereof. Justice processes which aim to address abuses committed in times of ongoing conflict in cases such as Sudan are now regularly conceptualised as transitional justice even if a transition is yet to occur – and it is uncertain if it will in any foreseeable future. This highlights a central expectation in contemporary transitional justice scholarship, namely that justice processes can help *initiate* a transition, rather than being pre-conditioned on the existence of it – or at least that what is now sometimes being referred to as 'pre-transition transitional justice' can advance other important goals such as providing victims with a level of redress.[65] However, in cases such as Sudan where a repressive regime remains in place and massive violations continue to take place there are obviously significant challenges related to operationalising any genuine justice process.[66] In other places, such as Colombia,[67] however, it has proven possible to bring in accountability measures in peace arrangements. Though frequently combined with some form of amnesty or reduced sentences, the move towards integrating accountability measures in peace agreements and other forms of settlements of armed conflict presents a significant development for how wars end.[68] This may impact the incentives of parties to a conflict in multiple ways, but does not necessarily make the conclusion of peace arrangements less likely.[69]

Challenges pursuing accountability for the crimes of major powers

Conflicts involving military presence by major powers create unique challenges from an accountability perspective, especially perhaps to the extent this occurs in fragile or (partly) collapsed States and involves military engagement with non-State actors. Sometimes, as in Iraq following the ousting of Saddam Hussein, such presence is authorised by the

---

[65]See, for example, Freedom House, *Delivering Justice Before and After Transitions*, 2013, available at https://freedomhouse.org/sites/default/files/Delivering%20Justice%20Before%20and%20After%20Transitions%20Istanbul%20Report%20Final%202014.pdf.

[66]See, for example, Brian Kritz and Jacqueline Wilson, 'No Transitional Justice without Transition: Darfur – A Case Study', *Michigan State Journal of International Law* 19/3 (2010–2011), 475–500.

[67]See, for example, Maria A. van Nievelt, 'Transitional Justice in Ongoing Conflict: Colombia's Integrative Approach to Peace and Justice', *Cornell International Affairs Review* 11/2 (2016), 101–138.

[68]The UN has stated that 'United Nations endorsed peace agreements can never promise amnesties for genocide, crimes against humanity or gross violations of human rights'. See UN Secretary-General, *The Rule of Law and Transitional Justice in Conflict and Post-Conflict Societies*, UN Doc. S/2004/616, 23 August 2004, para 10. On the actual use of amnesties, see further Louise Mallinder, *Amnesty, Human Rights and Political Transitions*.

[69]See further Mark Kersten, *Justice in Conflict*.

regime in place. In other cases, such as Syria, the regime has not consented to Western powers' military operations, thereby raising *jus ad bellum* questions in addition to the *jus in bello* issues discussed here. The type of crimes occurring in such conflicts tend to differ from the large-scale atrocities committed in the type of conflicts discussed above. Yet, some crimes, such as abuse of detainees, appear to have been committed systematically in Iraq, Afghanistan and other situations and were seemingly authorised by the military or political leaderships in the US and elsewhere.[70] Such situations pose particular challenges from an accountability perspective, both due to the nature of the abuses and the secrecy that often surround them and due to the ramifications of potentially implicating senior military and political leaders of major powers.

Whereas transitional justice has historically not been particularly occupied with abuses committed by major powers in the context of military campaigns and security operations abroad, accountability issues are now increasingly on the table, for a large part due to the activities of the ICC. Notably, since 2007, the ICC has conducted a preliminary examination of the situation in Afghanistan, involving scrutiny of the conduct of US military forces and the Central Intelligence Agency (CIA) relating to allegations of the war crimes of torture, rape and other crimes in detention facilities and in the context of renditions.[71] Emphasising the apparent systematic nature of these abuses and the general absence of domestic accountability processes for these crimes,[72] the ICC Prosecutor stated in November 2016 that the Office will 'imminently' make a final decision as to whether to request the Pre-Trial Chamber's authorisation of a formal investigation into the situation in Afghanistan.[73] Besides the preliminary examination in Afghanistan, the ICC Prosecutor is currently conducting preliminary examinations involving a number of other major powers (or their close allies), including an examination into the situation in Iraq, involving an assessment of whether British troops committed war crimes in detention centres and elsewhere during the Iraq war and occupation;[74] an examination of the situation in Palestine, involving an assessment of whether the Israeli Defence Forces committed war crimes

during the 2014 Gaza conflict and whether the Israeli government's settlement activities on West Bank territory amount to crimes under the

---

[70]For an account of US abuses in the war on terror, including a description of government authorisation under the Bush administration, see Human Rights Watch, 'Getting Away with Torture: The Bush Administration and Mistreatment of Detainees', 2011.
[71]See OTP Report on Preliminary Examination Activities (2016), paras. 198, 199 and 211.
[72]Ibid, paras. 212–220.
[73]Ibid, para 230.
[74]Ibid, paras. 75–108.

Court's jurisdiction;[75] and an examination of the situation in Ukraine, involving an assessment of the events in Crimea and Eastern Ukraine from 20 February 2014 onwards, assumedly including scrutiny of the conduct of pro-Russian forces.[76] The Office of the Prosecutor has further opened a full-scale investigation into the situation in Georgia relating to the August 2008 war between Georgia and Russia over the territory of South Ossetia involving allegations of crimes by the Russian armed forces.[77]

Taken together, these preliminary examinations and investigations suggest a significant shift in international justice whereby the conduct of major powers in armed conflict is increasingly being scrutinised by international justice institutions, specifically the ICC. This raises a range of novel and important questions. On the one hand, potential ICC prosecution of members of the armed and security forces of major powers, or even senior civil servants who may have authorised crimes, would present a major boost for accountability norms (and for the ICC as an institution). On the other hand, this would also bring the Court into a head-on collision with these powers, which it may not be geared to handle. It is, therefore, likely that the Office of the ICC Prosecutor views its institutional interests as best preserved if the opening of these preliminary examinations and investigations results that major powers undertake credible investigations and prosecutions domestically, in that way triggering the Court's so-called complementarity regime whereby ICC cases become inadmissible to the extent there are genuine domestic criminal processes dealing with the persons and incidents subject to ICC investigation.[78]

However, the extent to which so-called positive complementarity whereby the Court's activates are thought to encourage such genuine domestic processes actually 'works' is disputed. For example, the ICC's Iraq/UK preliminary examination has not 'triggered' a genuine domestic accountability process so far. Even if British officials have in the past cited to the ICC's preliminary examination as a justification for the continued existence of the Iraq Historic Allegations Team (IHAT) – a body set up to investigate allegations of crimes by the armed forces in Iraq – that body was created to satisfy the demands to investigate under human rights law, and in early 2017 the government announced its intention to dissolve IHAT,

---

[75]Ibid, paras. 109–145.

[76]Ibid, paras. 146–191.

[77]See Office of the Prosecutor of the International Criminal Court, *Request for Authorisation of an Investigation Pursuant to Article 15*, ICC-01/15–4, 13 October 2015, paras. 98; 140.

[78]For a detailed account of such strategies in the Iraq/UK preliminary examination, see further Thomas Obel Hansen, 'Accountability for British War Crimes in Iraq? Examining the Nexus between International and National Justice Responses', in Morten Bergsmo and Carsten Stahn (eds.), *Quality Control in Preliminary Examination: Reviewing Impact, Policies and Practices* (TOAEP 2017), 399–450.

notwithstanding that the ICC's preliminary examination is still ongoing.[79] This suggests that the ICC may have a less decisive impact on decision-making processes relating to domestic justice processes in the UK than hoped for by advocates of positive complementarity.[80] There are even fewer reasons to believe that the US – which is not a State Party to the Rome Statute and has in the past taken a hostile attitude towards the ICC when perceiving the Court's actions to contravene its interests – will fundamentally alter its approach to accountability for torture and other war crimes allegedly committed in Afghanistan and elsewhere as a consequence of ICC activities.[81]

Despite the activities of the ICC, there are significant challenges related to promoting criminal accountability for members of the political and military leadership of major powers who allegedly authorised or accepted the use of methods of warfare in violation of international law. However, other less far-reaching approaches to accountability may prove more feasible addressing these types of abuses. For example, lawyers have successfully brought civil suits against the Ministry of Defense in the UK, leading to settlements whereby hundreds of Iraqi victims have received compensation.[82] The UK has also witnessed a continued debate about the legality of the Iraq war, including attempts at holding to account the political leaders who ordered and planned the invasion.[83]

Conclusions

Justice tools, regularly conceptualised as transitional justice, are increasingly applied to situations of armed conflict, both after war has ended and while hostilities are still ongoing. Transitional justice claims to possess the ability to both contribute to resolving and regulating the conduct in armed conflict. Exploring these claims, this article has pointed to the significant progress made giving effect to accountability norms for crimes committed in

---

[79]Ibid.

[80]For examples of such expectations, see, for example, Fatou Bensouda, 'Reflections from the International Criminal Court Prosecutor', *Case Western Reserve Journal of International Law* 45 (2012), at 508–509.

[81]On US-ICC relations, see further David Bosco, *Rough Justice: The International Criminal Court in a World of Power Politics* (Oxford: OUP 2014).

[82]'Hundreds of compensation claims against British soldiers could be abandoned after controversial law firm announces closure', *The Telegraph*, 15 August 2016, available at http://www.telegraph.co.uk/news/2016/08/14/hundreds-of-compensation-claims-against-british-soldiers-could-b/.

[83]Lawsuits relating to the legality of the 2003 Iraq War, brought on the basis of rules concerning private prosecution, have been filed in British courts, requesting trial of then prime minister Tony Blair, foreign secretary Jack Straw, and Lord Goldsmith, the attorney general at the time. However, the High Court ruled in July 2017 that the crime of aggression does not exist under English law and hence blocked the suit. See further 'Tony Blair Prosecution over Iraq War blocked by Judges', *The Guardian*, 31 July 2017, available at https://www.theguardian.com/politics/2017/jul/31/tony-blair-prosecution-over-iraq-war-blocked-by-judges.

armed conflict, but also a range of complexities and challenges in so doing. Concerning the ability of transitional justice to contribute to ending war, it is clear that the correlations between peace and justice cannot be summarised in simple slogans such as 'peace versus justice' or 'no peace without justice'. Rather, justice processes entail a range of dynamics which impact peace processes in multiple, sometimes contradicting ways, and which can vary significantly over time and space. Concerning the ability of transitional justice to regulate the conduct of parties to armed conflict, the nature of contemporary forms of conflict presents significant challenges, for example due to the magnitude of abuses frequently associated with civil wars and because a distinction between victims and perpetrators cannot always be easily drawn. Furthermore, justice for atrocities in war, whether occurring at the national or international level, continues to be usually one-sided. Unsurprisingly, it has in particular proven problematic pursuing political and military leaders who continue to hold power. At the same time, however, accountability for crimes committed by major powers in armed conflict is now increasingly on the agenda, in particular due to the ICC's recent activities.

Disclosure statement

No potential conflict of interest was reported by the author.

Bibliography

Arthur, Paige, 'How "Transitions" Reshaped Human Rights: A Conceptual History of Transitional Justice', *Human Rights Quarterly* 31/2 (2009), 321–67. doi:10.1353/hrq.0.0069

Backer, David, 'Cross-National Comparative Analysis', in Hugo van der Merwe, Victoria Baxter, and Audrey Chapman, (eds.), *Assessing the Impact of Transitional Justice* (Washington DC: United States Institute of Peace Studies 2009) 23–90.

Barnes, Gwen P., 'The International Criminal Court's Ineffective Enforcement Mechanisms: The Indictment of President Omar Al Bashir', *Fordham International Law Journal* 34/6 (2011), 1585–619.

Bassiouni, M. Cherif, 'Accountability for Violations of International Humanitarian Law and Other Serious Violations of Human Rights', in Bassiouni, (ed.), *Post-Conflict Justice* (New York: Transnational Publishers 2002) 3–54.

Bensouda, Fatou, 'Reflections from the International Criminal Court Prosecutor', *Case Western Reserve Journal of International Law* 45 (2012), 505–11.

Bosco, David, *Rough Justice: The International Criminal Court in a World of Power Politics* (Oxford: OUP 2014).

Clark, Phil, 'Establishing a Conceptual Framework: Six Key Transitional Justice Themes', in Phil Clark and Zachary Kaufman, (eds.), *After Genocide: Transitional Justice, Post-Conflict Reconstruction and Reconciliation in Rwanda and Beyond* (London: Hurst 2008) 191–205.

Collins, Cath, 'The End of Impunity? "Late Justice" and Post-Transitional Justice in Latin America', in Palmer, et al., (ed.), *Critical Perspectives in Transitional Justice* (Cambridge: Intersentia 2012), 399–423.

deGuzman, Margaret M., 'Choosing to Prosecute: Expressive Selection at the International Criminal Court', *Michigan Journal of International Law* 33/2 (2012), 265–320.

Drumbl, Mark, 'The Ongwen Trial at the ICC: Tough Questions on Child Soldiers', openDemocracy, 14 April 2015, available at https://www.opendemocracy.net/openglobalrights/mark-drumbl/ongwen-trial-at-icc-tough-questions-on-child-soldiers

Dutton, Yvonne M., 'Enforcing the Rome Statute: Evidence of (Non) Compliance from Kenya', *Indiana International and Comparative Law Review* 26/7 (2016), 7–32. doi:10.18060/7909.0034

Ellis, Mark, 'Combating Impunity and Enforcing Accountability as a Way to Promote Peace and Stability – The Role of International War Crimes Tribunals', *Journal of National Security and Policy* 2/1 (2006), 111–64.

Engstrom, Par, 'Transitional Justice and Ongoing Conflict', in Chandra Lekha Sriram, et al., (ed.), *Transitional Justice and Peacebuilding on the Ground: Victims and Excombatants* (London: Routledge 2012), 41–61.

Freedom House, Delivering Justice before and after Transitions, 2013, available at https://freedomhouse.org/sites/default/files/Delivering%20Justice%20Before%20and%20After%20Transitions%20Istanbul%20Report%20Final%202014.pdf

Hansen, Thomas Obel, 'Transitional Justice in Kenya? an Assessment of the Accountability Process in Light of Domestic Politics and Security Concerns', *California Western International Law Journal* 42/1 (2011), 1–35.

Hansen, Thomas Obel, 'Transitional Justice: Toward a Differentiated Theory', *Oregon Review of International Law* 13/1 (2011), 1–46.

Hansen, Thomas Obel, 'The International Criminal Court in Kenya: Three Defining Features of a Contested Accountability Process and Their Implications for the Future of International Justice', *Australian Journal of Human Rights* 18/2 (2012), 187–217. doi:10.1080/1323-238X.2012.11882112

Hansen, Thomas Obel, 'The Vertical and Horizontal Expansion of Transitional Justice: Explanations and Implications for a Contested Field', in Susanne Buckley-Zistel, et al., (ed.), *Transitional Justice Theories* (London: Routledge 2013), 105–24.

Hansen, Thomas Obel, 'The International Criminal Court and the Legitimacy of Exercise', in Per Andersen, et al., (ed.), *Law and Legitimacy* (Copenhagen: DJOEF 2015), 73–100.

Hansen, Thomas Obel, 'The Time and Space of Transitional Justice', in Dov Jacobs, et al., (ed.), *Research Handbook on Transitional Justice* (Cheltenham: Edward Elgar Publishing 2017), 34–51.

Hansen, Thomas Obel, 'Accountability for British War Crimes in Iraq? Examining the Nexus between International and National Justice Responses', in Morten Bergsmo and Carsten Stahn, (eds.), *Quality Control in Preliminary Examination: Reviewing Impact, Policies and Practices* (Florence, Italy: TOAEP 2018), 399–450.

Human Rights Watch, *Getting Away with Torture: The Bush Administration and Mistreatment of Detainees* (New York, NY 2011).

Huntington, Samuel, *The Third Wave: Democratization in the Late Twentieth Century* (Oklahoma: University of Oklahoma Press 1991).

Iverson, Jens, 'Transitional Justice, Jus Post Bellum and International Criminal Law: Differentiating the Usages, History and Dynamics', *International Journal of Transitional Justice* 7/3 (2013), 413–33. doi:10.1093/ijtj/ijt019

Jo, Hyeran and Beth A. Simmons, 'Can the International Criminal Court Deter Atrocity?', *International Organization* 70/3 (2016), 443–75. doi:10.1017/S0020818316000114

Kersten, Mark, *Justice in Conflict: The Effects of the International Criminal Court's Interventions on Ending Wars and Building Peace* (Oxford: OUP 2016).

Kritz, Brian and Jacqueline Wilson, 'No Transitional Justice without Transition: Darfur – A Case Study', *Michigan State Journal of International Law* 19/3 (2010–2011), 475–500.

Kritz, Neil J., (ed.), *Transitional Justice: How Emerging Democracies Reckon with Former Regimes*, Volume Vol. I and II (Washington DC: United States Institute of Peace Press 1995).

Labuda, Patryk, 'Taking Complementarity Seriously: Why Is the International Criminal Court Not Investigating Government Crimes in Congo?', *Opinio Juris*, 28 April 2017, available at http://opiniojuris.org/2017/04/28/33093/?utm_source=feedburner&utm_medium=email&utm_campaign=Feed%3A+opiniojurisfeed+%28Opinio+Juris%29

Lemarchand, René, 'The Politics of Memory in Post-Genocide Rwanda', in Phil Clark and Zachary Kaufman, (eds.), *After Genocide: Transitional Justice, Post-Conflict Reconstruction and Reconciliation in Rwanda and Beyond* (London: Hurst Publishers 2008), 65–76.

Linz, Juan J. and Alfred Stepan, *Problems of Democratic Transition and Consolidation: Southern Europe, South America, and Post-Communist Europe* (Maryland: The Johns Hopkins University Press 1996).

Mallinder, Louise, *Amnesty, Human Rights and Political Transitions: Bridging the Peace and Justice Divide* (Oxford: Hart Publishing 2008).

Moffett, Luke, 'Elaborating Justice for Victims at the International Criminal Court', *Journal of International Criminal Justice* 13 (2015), 281–311. doi:10.1093/jicj/mqv001

Moreno-Ocampo, Luis, 'Transitional Justice in Ongoing Conflicts', *International Journal of Transitional Justice* 1/1 (2007), 8–9. doi:10.1093/ijtj/ijm014

Murray, Daragh, *Practitioners' Guide to Human Rights Law in Armed Conflict* (Oxford: OUP 2016).

Mutyaba, Rita, 'An Analysis of the Cooperation Regime of the International Criminal Court and Its Effectiveness in the Court's Objective in Securing Suspects in Its Ongoing Investigations and Prosecutions', *International Criminal Law Review* 12 (2012), 937–62. doi:10.1163/15718123-01205006b

Nagy, Rosemary, 'Transitional Justice as Global Project: Critical Reflections', *Third World Quarterly* 29/2 (2008), 275–89. doi:10.1080/01436590701806848

Ni Aoláin, Fionnuala and Colm Campbell, 'The Paradox of Transition in Conflicted Democracies', *Human Rights Quarterly* 27 (2005), 172–213. doi:10.1353/hrq.2005.0001

Nino, Carlos, 'Response: The Duty to Punish past Abuses of Human Rights into Context: The Case of Argentina', in Neil J Kritz, (ed.), *Transitional Justice: How Emerging Democracies Reckon with Former Regimes*, Volume Vol. I (Washington DC: United States Institute of Peace Press 1995), 417–36.

Nouwen, Sarah and Wouter Werner, 'Doing Justice to the Political: The International Criminal Court in Uganda and Sudan', *The European Journal of International Law* 21/4 (2011), 941–65. doi:10.1093/ejil/chq064

O'Donnell, Guillermo and Philippe Schmitter, *Transitions from Authoritarian Rule: Tentative Conclusions about Uncertain Democracies* (Maryland: The Johns Hopkins University Press 1986).

Ogora, Lino Owor, 'Just or Unjust: Mixed Reactions on whether Ongwen Should Be on Trial', *JusticeHub*, 25 April 2017, available at https://justicehub.org/article/just-or-unjust-mixed-reactions-whether-ongwen-should-be-trial

Orentlicher, Diane F., 'A Reply to Professor Nino', in Neil J Kritz, (ed.), *Transitional Justice: How Emerging Democracies Reckon with Former Regimes*, Volume Vol. I (Washington DC: United States Institute of Peace Press 1995), 2641–43.

Peskin, Victor, 'Beyond Victor's Justice? the Challenge of Prosecuting the Winners at the International Criminal Tribunals for the Former Yugoslavia and Rwanda', *Journal of Human Rights* 4/2 (2005), 213–31. doi:10.1080/14754830590952152

Posner, Eric and Adrian Vermeule, 'Transitional Justice as Ordinary Justice', *Harvard Law Review* 117/762 (2003), 762–825.

Reiter, Andrew, et al., 'Transitional Justice and Civil War: Exploring New Pathways, Challenging Old Guideposts', *Transitional Justice Review* 1/1 (2012), 137–69.

Reynolds, Abigail, 'Deterring the Use of Child Soldiers in Africa: Addressing the Gap between the Mandate of the International Criminal Court and Social Norms and Local Understandings', master thesis submitted to Leiden University, June 2016, available at https://openaccess.leidenuniv.nl/bitstream/handle/1887/41163/Reynolds,%20Abigail-s1754807-MA%20Thesis%20PS-2016.pdf?sequence=1

Roht-Arriaza, Naomi, 'The New Landscape of Transitional Justice', in Naomi Roht-Arriaza and Javier Mariezcurrena, (eds.), *Transitional Justice in the Twenty-First Century: Beyond Truth versus Justice* (Cambridge: CUP 2006) 1–16.

Schabas, William A., 'The Rwanda Case: Sometimes It's Impossible', in M. Cherif Bassiouni, (ed.), *Post Conflict Justice* (New York: Transnational 2002) 499–522.

Schabas, William A., 'State Policy as an Element of International Crimes', *Journal of Criminal Law and Criminology* 98 (2008), 953–82.

Sharp, Dustin, 'Beyond the Post-Conflict Checklist: Linking Peacebuilding and Transitional Justice through the Lens of Critique', *Chicago Journal of International Law* 14 (Summer 2013), 165–96.

Sharp, Dustin, 'Interrogating the Peripheries; the Preoccupations of Fourth Generation Transitional Justice', *Harvard Human Rights Journal* 26 (2013), 149–78.

Sikkink, Kathryn, *The Justice Cascade: How Human Rights Prosecutions are Changing World Politics* (New York and London: W.W. Norton and Company 2011).

Snyder, Jack and Leslie Vinjamuri, 'To Prevent Atrocities, Count on Politics First, Law Later', openDemocracy, 12 May 2015, available at https://www.opendemocracy.

net/openglobalrights/jack-snyder-leslie-vinjamuri/to-prevent-atrocities-count-on-politics-first-law-late

Sriram, Chandra Lekha, 'Justice as Peace? Liberal Peacebuilding and Strategies of Transitional Justice', *Global Society* 21/4 (2007), 579–91. doi:10.1080/13600820701562843

Sriram, Chandra Lekha, Johanna Herman, and Olga Martin-Ortega, 'Beyond Justice versus Peace: Transitional Justice and Peacebuilding Strategies', in Karin Aggestam and Annika Björkdahl, (eds.), *Rethinking Peacebuilding: The Quest for Just Peace in the Middle East and the Western Balkans* (New York: Routledge 2012) 1–23.

Stahn, Carsten, "Jus Ad Bellum', 'Jus in Bello'.. 'Jus Post Bellum'? – Rethinking the Conception of the Law of Armed Force', *The European Journal of International Law* 17/5 (2007), 921–43. doi:10.1093/ejil/chl037

Teitel, Ruti, *Transitional Justice* (Oxford: OUP 2000).

Unger, Thomas and Marieke Wierda:, 'Pursuing Justice in Ongoing Conflict: A Discussion of Current Practice', in K. Ambos, J. Large, M. Wierda (ed.), *Building a Future on Peace and Justice* (Berlin: Springer 2009), 263–302.

van Den Wyngaert, Christine, 'Victims before International Criminal Courts: Some Views and Concerns of an ICC Trial Judge', *Case Western Reserve Journal of International Law* 44 (2011), 475–96.

van der Merwe, Hugo, Victoria Baxter, and Audrey Chapman, (eds.), *Assessing the Impact of Transitional Justice* (Washington DC: United States Institute of Peace Studies 2009).

van Nievelt, Maria A., 'Transitional Justice in Ongoing Conflict: Colombia's Integrative Approach to Peace and Justice', *Cornell International Affairs Review* 11/2 (2016), 101–38.

Vinjamuri, Leslie, 'Trials and Errors: Principle and Pragmatism in Strategies of International Justice', *International Security* 28 (2003), 5–44.

Zalaquett, José, 'Balancing Ethical Imperatives and Political Constraints: The Dilemma of New Democracies Confronting past Human Rights Violations', in Neil J. Kritz, (ed.), *Transitional Justice: How Emerging Democracies Reckon with Former Regimes*, Volume Vol. I (Washington DC: United States Institute of Peace Press 1995), 203–06.

Zalaquett, José, 'Confronting Human Rights Violations Committed by Former Governments: Principles Applicable and Political Constraints', in Neil J. Kritz, (ed.), *Transitional Justice: How Emerging Democracies Reckon with Former Regimes*, Volume Vol. I (Washington DC: United States Institute of Peace Press 1995), 3–31.

The Joint Chiefs of Staff, the atom bomb, the American military mind and the end of the Second World War

Phillips Payson O'Brien

ABSTRACT

The decision by the US government to drop the atomic bombs on Japan is one of the most heavily debated questions in history. This article examines one element of that debate, in many ways the most surprising. That was the different views of the top of the military hierarchy in the USA, the Joint Chiefs of Staff (JCS). The JCS was on the whole more sceptical about using atomic weaponry than the USA's civilian leadership, for ethical and strategic reasons. As such they were willing to consider very different ways of ending the war.

The Second World War witnessed the most famous and controversial ending of any conflict in human history. The decision by the US government to drop uranium and plutonium, fission-based weapons (better known as atomic bombs) on the Japanese cities of Hiroshima and Nagasaki makes its ending seminal. From the moment the bombs exploded there has been a debate, which shows no sign of abating, about the decision to cross the atomic threshold; whether the act was necessary, ethical or even counter-productive. A multitude of books and articles have been written on the subject that go back and forth over the question of whether this was the best way to achieve victory.[1]

---

[1]Much of the vast literature which delves into the Truman administration's decision to use the atom bomb, falls into three large groupings. The traditionalists/orthodox historians argue that the atom bomb was needed to force Japanese surrender, prevent a bloody invasion of Japan, and save many tens, if not hundreds of thousands of US and Japanese lives. Some of the best examples include: Wilson Miscamble, *The Most Controversial Decision: Truman, the Atomic Bombs and the Defeat of Japan* (Cambridge: Cambridge Univ Press 2011); D. M. Giangreco, *Hell to Pay: Operation Downfall and the Invasion of Japan 1945–1947* (Annapolis: Naval Institute Press 2009); Robert James Maddox, *Weapons for Victory: The Hiroshima Decision* (Columbia: Univ Missouri Press 2004); Richard B. Frank, *Downfall: the End of the Imperial Japanese Empire* (New York: Penguin 2001). Opposed to this group are the revisionists. They argue that the bomb did not need to be dropped as Japan was already close to surrender and that any US invasion of japan was not

This article will not rehash those arguments. What it aims to do, however, is examine one particular set of views, in many ways the most surprising – that of the top of the US strategic command, the Joint Chiefs of Staff (JCS). This group of two army generals and two navy admirals, all profoundly conventional, Caucasian, Christian, educated in military academies, born and raised in middle-class households – had a more complex set of ethical and strategic responses to the use of atomic weapons than one might assume. Looking at the range of their views on the issue of the atom bomb provides an important insight into how they viewed victory, the meaning of the Second World War and the state of the American Military Mind.

The JCS were the four most important military officers in the USA during the Second World War: Admiral William Leahy, Chief of Staff to the President and Chairman of the joint chiefs; General George C. Marshall, Chief of Staff of the Army; Admiral Ernest King, Chief of Naval Operations; General Henry (Hap) Arnold, Chief of Staff of the Army Air Force.[2] Collectively, they were Franklin Roosevelt's and Harry Truman's highest ranking military advisers. Leahy was one of Roosevelt's oldest friends in Washington; the two had become close in 1913 when they first met in the Navy Department. Marshall, King and Arnold had more professional relationships with the president and this made a material difference over the eventual decision to use the atomic bomb. While Roosevelt lived, Leahy had great influence over the president, particularly as time went on and Roosevelt weakened. After Roosevelt's death in April 1945, however, things changed. Truman was an army man by background and esteemed George Marshall in particular. The shift would prove to be important as Leahy and Marshall's views on the atom bomb diverged the most as time went by.

---

necessary or if it had to go ahead that it would have resulted in far fewer casualties than the traditionalists argue, and that the bomb was either mostly or partially aimed at intimidating the USSR. Some of the best examples include: Campbell Craig and Sergey Radchenko, *The Atomic Bomb and the Origins of the Cold War* (New Haven: Yale Univ Press 2008); Kai Bird and Lawrence Lifschultz (eds.), *Hiroshima's Shadow* (Stony Creek: The Pamphleteers Press 1998); Gar Alperovitz et al., *The Decision to Use the Atomic Bomb: And the Architecture of an American Myth* (New York: Alfred Knopf 1995); Gar Alperovitz, *Atomic Diplomacy: Hiroshima and Potsdam, The Use of the Atomic Bomb and American Confrontation with Soviet Power* (New York: Vintage 1965). There is a middle ground that argues that the bomb was really dropped to try and force the Japanese to surrender, not to intimidate the Soviets, but that the Japanese probably would have surrendered relatively quickly anyway, certainly not long after the Red Army attacked. For some of these analyses, see J. Samuel Walker, *Prompt and Utter Destruction: Truman and the Use of Atomic Bombs against Japan* (Chapel Hill: UNC Press 1997); Kenneth P. Werrell, *Blankets of Fire: US Bombers over Japan during World War II* (Washington: Smithsonian Institute Press 1996). Walker has also published a very useful historiographical article on the different schools of thoughts: J. Samuel Walker, 'Recent Literature on Truman's Atomic Bomb Decision', in Frank Costigliola and Michael J, Hogan (eds.), *America in the World: The Historiography of American Foreign Relations since 1941* (Cambridge: Cambridge Univ Press 2014). Walker does an excellent job summarising the different arguments on the bomb and shows the strength of a middle ground position between the more hard line traditionalists and revisionists. For an overview of some of the ethical views of the dropping of the bomb, see Francis X. Winters, *Remembering Hiroshima: Was it Just* (Farnham: Ashgate 2009).

[2]One of the best books about the Joint Chiefs is: Mark A. Stoler, *Allies and Adversaries, The Joint Chiefs of Staff, the Grand Alliance, and US Strategy in World War II* (Chapel Hill: UNC Press 2000). Stoler spends only a little time discussing the different joint chiefs and the decision to use the atomic bomb (256–7).

The other two joint chiefs, King and Arnold, were less involved in the ultimate decision of whether and how to use the atom bomb. King was handicapped by the fact that the development of the atom bomb through the Manhattan Project was an Army programme.[3] As such Arnold, who as the head of the air force, though actually still under Marshall who was head of the army, could have tried to play an important role in determining atomic policy, but was somewhat passive on the issue. His reluctance to get involved came at least partly from his poor health (Arnold had a number of major heart attacks during the war).

The first thing that must be understood about the joint chiefs was that they were less important than the civilian leadership in making the final decision to drop the atom bomb. While Roosevelt was alive, he kept such important decisions very close to his chest, often deciding things using only the joint chiefs with Harry Hopkins. Truman, on the other hand, feeling a little unsure and wanting very much to act in what he thought was Roosevelt's spirit (but which decidedly was not) delegated a great deal of authority in the first few months of his Interim Committee to decide on the best way forward.[4] Composed of government officials and scientists, the committee was chaired by Secretary of War Henry Stimson and, crucially, included James Byrnes as Truman's personal representative. For Stimson, this represented a temporary increase in influence. While Roosevelt lived, the president had kept the elderly Republican away from many of the crucial meetings of the war. Stimson had, for instance, not been included in the grand strategic conferences in Casablanca, Quebec (both first and second), Cairo, Tehran and Yalta – conferences which determined the strategy of the war and set many of conditions that were supposed to govern the peace afterwards. Truman, however, allowed the secretary of war to have some influence over important decisions such as the atom bomb.[5] However even more than Stimson, the real beneficiary of the Interim Committee's establishment was Byrnes. Roosevelt had been increasingly irritated with Byrnes and just before FDR died, the South Carolinian was leaving his important position as head of the Office of Economic Stabilization. Truman, however, leaned heavily on Byrnes, whom he had originally thought Roosevelt would select to be the new vice president in 1944. Not only did he make Byrnes his personal representative on the Interim Committee, in late June 1945 he would

---

[3]Ernest J. King and Walter Muir Mitchell, *Fleet Admiral King: A Naval Record* (London: Eyre and Spottiswoode 1953), 411–12.

[4]Harry S. Truman, *Memoirs: Volume One, Year of Decisions* (Garden City: Doubleday 1955), 418.

[5]Truman actually invited Stimson to attend the Potsdam Conference in July 1945, but by the time the meetings started he had decided he could do without the Secretary of War's advice and kept Stimson off the official American negotiating team.

soon make him the new secretary of state, when the incumbent, Edward Stettinius, resigned.

The influence of Stimson and Byrnes mattered greatly as the former was mostly supportive of using the atomic bomb while the latter was enthusiastically supportive, and this made the formation of the Interim Committee vital. During the meeting of the committee on 31 May 1945, Stimson, who was honestly torn about the meaning of the bomb, endorsed the use of the atomic bomb against Japan without warning.[6] However, he made a contradictory claim that while the attack should not 'concentrate' on a civilian area, an area of war industry closely surrounded by workers houses was the most 'desirable target'.[7] Byrnes had no such qualms. When the committee met again on 1 June, he dispensed with any of Stimson's qualifications. According to the minutes:

> 'VI Use Of The Bomb:
>
> Mr. Byrnes recommended, and the committee agreed, that the Secretary of War should be advised that, while recognizing that the final selection of the target was essentially a military decision, the present view of the Committee was that the bomb should be used against Japan as soon as possible; that it be used on a war plant surrounded by workers' homes; and that it be used without prior warning'.[8]

One of the reasons Byrnes felt free to speak so forcefully on the issue is that he would have known that Truman, instinctively, was also supportive of using the bomb against Japan as soon as it was ready. Unlike Roosevelt, who seemed to become more and more enigmatic about the atomic bomb's use the closer it came to being a reality,[9] Truman was more openly supportive about dropping the bomb and even seemed slightly energised by the prospect of using the new weapon. He made no ethical comments about the weapon, except just before it was used

---

[6]Sean L. Malloy, *Atomic Tragedy: Henry L. Stimson and the Decision to Use the Atomic Bomb against Japan* (Ithaca: Cornell Univ Press 2018), 115–16. Malloy described Stimson's conflicts over the bomb in detail, and argues that ultimately he accepted that it could be used on large Japanese city (as long as it was not Kyoto) partly because the shock value could lead to a Japanese surrender. According to one of Stimson's biographers, the Secretary of War always assumed the bomb would be dropped, so never really contemplated holding back. Elting E. Morison, *Turmoil and Tradition: A Study of the Life and Times of Henry L. Stimson* (Boston: Houghton Mifflin 1960), 629.

[7]'Notes of the Meeting of the Interim Committee, 31 May 1945', pp 13–14. Accessed online through the Truman Library: https://www.trumanlibrary.org/whistlestop/study_collections/bomb/large/docu ments/index.php?documentid=39&pagenumber=14.

[8]'Notes of the Meeting of the Interim Committee, 1 June 1945', pp. 8–9. Accessed online through the Truman Library: https://www.trumanlibrary.org/whistlestop/study_collections/bomb/large/docu ments/index.php?documentid=40&pagenumber=8.

[9]Roosevelt said very little about the bomb the closer it came to being a reality. After his meetings with Churchill in October 1944, he was reluctant to discuss its use. One of the few things he said about it was that it might be used against Japan, but only in a demonstration over an uninhabited area. See Ronald Takaki, *Hiroshima: Why America Dropped the Bomb* (Boston: Little Brown 1995), 20–1; Alperovitz, *The Decision to Use the Atomic Bomb*, 662.

when he wrote in a diary that it should not be used against women and children.[10] However, he made no order to that effect, and when the first bomb was dropped on Hiroshima, he celebrated with gusto.[11]

Thus, the civilian leadership was far more unified in favour of using the weapon and the civilian leadership, led by Truman, ultimately gave the go ahead for its use. This is particularly important because, had it been up to the US military led by the joint chiefs, the atom bomb might never have been used. The truth of the matter was that Leahy, Marshall, King and Arnold were all, to some degree, hesitant to use the atom bomb.[12] The chiefs, and many other senior military personalities such as generals Dwight Eisenhower and Douglas MacArthur and admirals Chester Nimitz and William Halsey all argued, mostly after the war it must be said, that dropping the atom bomb was either unnecessary or wrong.[13] Within the chiefs there was a realisation that the dropping of the bomb could shape the victory and the post-war world.

King and Arnold: the soft middle

Though all the chiefs had some doubts, they approached the issue of the use of the atom bomb and the end of the war from quite different perspectives. The two most difficult to pin down, partly because their post-war stance conflicted with the contemporary record from May to August 1945, were King and Arnold. Both made negative comments about the atom bomb after the war, but there is little, if any, indication that they either held those views or tried to put them into action while the war was ongoing.

Ernest King was the most enigmatic. There is no useful record of King saying anything about the atomic bomb before Hiroshima and Nagasaki were attacked, though he first was given information about the Manhattan Project in 1943.[14] He never attended the Interim Committee and there seem

---

[10]This was in a series of pages of notes that Truman kept during Potsdam. The entry about not attacking women and children was dated 25 July 1945. The document can be accessed at: http://www.trumanlibrary.org/whistlestop/study_collections/bomb/large/documents/pdfs/63.pdf#zoom=100.

[11]Arnold A. Offner, *Another Such Victory: President Truman and the Cold War 1945–1953* (Palo Alto: Stanford Univ Press 2002), 92.

[12]In: Alperovitz, *The Decision to Use the Atomic Bomb* (1995), there are large sections on all the different military personalities that expressed some opposition to the use of the atomic bomb. It is useful, but at times lacks nuance as it ascribes a rather one dimensional nature to their thinking. See pages 329, 334–5, 348 for some mentions of King and Arnold. From these, one would assume that they were more clearly opposed to the use of atomic weapons than they probably were.

[13]Eisenhower, Nimitz and Halsey all spoke negatively about the use of the atom bomb, in an ethical sense. They not only believed that the weapon was unnecessary to end the war, they also believed that it was morally wrong for the USA to have used the weapon in the first place.

[14]Thomas B. Buell, *Master of Sea Power: A Biography of Fleet Admiral Ernest J. King* (Annapolis: Naval Institute Press 1980), 496.

to be no minutes kept of a JCS meeting during which the bomb was discussed.[15] After the war, King was only slightly more forthcoming. He was quoted not long after the war saying that he did not like 'any part of' the atomic weapon.[16] This has been used by a number of scholars to claim King was opposed to the attacks.[17] However, this claim rests on very light evidence. It is probably best to say that King had doubts about the bomb, but was happy to stay out of the discussions and as such tacitly gave consent. In his private papers, there are pages and pages of notes that he assembled to help write a memoir. They have stories ranging from the most mundane to the highest levels of grand strategy in the war.[18] Yet, in none of them does he address the question of the atom bomb in any detail. Only in the memoir itself, which he wrote (in the third person) with the assistance of a reserve naval officer, did King provide any meaningful glimpse into his thinking – and in an extremely short meditation. He was very negative about the use of the bomb, though more on the grounds of necessity than ethics.

> The President in giving his approval for these attacks appeared to believe that many of thousands of American troops would be killed in invading Japan, and in this he was entirely correct; but King felt, as he had pointed out many times, that the dilemma was unnecessary one, for had we been willing to wait, the effective naval blockade would, in the course of time, have starved the Japanese into submission through lack of oil, rice, medicines, and other essential materials. The Army, however, with its underestimation of sea power, had insisted upon a direct invasion and an occupational conquest of Japan proper. King still believes this was wrong.[19]

Of all the statements made by members of the joint chiefs about their general view of the atom bomb and the end of the war against Japan, this might be the most problematic. King is absolutely right in pinpointing that the discussion over the atom bomb cannot be seen in isolation, it was part of a larger debate about whether Japan should be invaded or not. The atom bomb itself was not successfully tested until 16 July, and until then there was a real doubt as to whether the weapon would work and even if it did, how destructive it would turn out to be. As such, before 16 July, there was an ongoing discussion over a possible invasion of Japan by US ground

---

[15]For instance, between 1 March 1945 and the attack on Hiroshima, which would have been the period if any during which the atomic bomb would have been mentioned, there is no minuted discussion of it in any of the JCS minutes. During that period the JCS met on: 1 March, 13 March, 27 March, 24 April, 22 May, 12 June, 18 June, 26 June, 16–21 July (as part of Potsdam Conference), 23 July.

[16]Buell, *Master of Sea Power*, 497.

[17]Ronald Schaffer, *Wings of Judgement: American Bombing in World War II* (New York: Oxford Univ Press 1985), 166; Alperovitz, *The Decision to Use the Atomic Bomb*, 329.

[18]Ernest King Mss, Library of Congress Manuscript Division, Washington, DC, Box 35 has over 100 pages of different notes that King put together after the war. There does not seem to be a mention of the atom bomb anywhere in them.

[19]King and Whitehill, *Fleet Admiral King, A Naval Record* (London 1953), 412.

forces, starting with the main southern island of Kyushu. This invasion talk complicates the ethical strategic discussion, as any invasion held out the possibility of high casualties amongst both US and Japanese military personnel. It also held out the possibility of a high number of Japanese civilian deaths, even more than occurred during the bombing of Hiroshima and Nagasaki.[20]

However, King's claim that he had consistently argued against any invasion of Japan is deceptive. He might have opposed an invasion of Honshu, Japan's largest island, but he made a strong case for the invasion of the southernmost island, Kyushu. In June 1945, when discussion over the invasion reached a fever pitch, he spoke clearly in support of an assault on Kyushu.[21] There was a famous meeting in the White House on 18 June involving Truman, the joint chiefs (though Arnold could not make it and was represented by General Ira Eaker), Stimson and the secretary of the navy, James Forrestal. During this meeting, King spoke strongly in favour of an invasion of Kyushu, indeed, he sided with Marshall who painted a generally rosy picture of the casualties the USA would suffer in any such operation. King also claimed that seizing Kyushu would provide the USA with real strategic advantages.

> Admiral King agreed with General Marshall's views and said that the more he studied the matter, the more he was impressed with the strategic location of Kyushu, which he considered the key to success of any siege operations.... It was his opinion that we should do Kyushu now, after which there would be time to judge the effect of possible operations by the Russians and the Chinese.[22]

King's post-war comments about his supposed reluctance to use either the bomb or invade Japan, therefore, need to be taken with a large grain of salt. They are an indication that he was probably the least ethically engaged member of the JCS when it came to the use of the atom bomb and the end of the war against Japan. The opposition that he mentioned after the war, instead, had everything to do with proving the truth of his vision about the importance of sea power. It says a lot about the person. King fought the entire war trying to show the importance (and independence) of the US Navy, as such he was often opposed to cooperating closely with either the

---

[20]The weakest argument made by the revisionists against the atomic bomb was that the casualties that would have been incurred during any such operation would have been on the lighter side: (see Barton Berstein, 'A Postwar Myth: 500,000 US Lives Saved', *Bulletin of the Atomic Scientists* 42, (June/July 1986), 38–40; John R. Skates, *The Invasion of Japan, Alternative to the Bomb* (Columbia: Univ South Carolina Press 1994), 76–80). It was also an unnecessary diversion from the most important ethical argument against dropping the bomb, not that it precluded any invasion, but that no invasion was necessary because of US air and sea control around Japan.

[21]The minutes of this meeting are discussed in every serious work on the debate over invading Japan or using the atom bomb. For an online copy, see 'Minutes of Meeting held at the White House, 18 June, 1945', Accessed online through the Truman Library. https://www.trumanlibrary.org/whistle stop/study_collections/bomb/large/documents/index.php?documentid=21&pagenumber=1.

[22]Ibid, 4.

American Army or the British Navy. He was worried, at the end of the war, that the atom bomb would overshadow what he believed was the US Navy's decisive contribution to victory in the war against Japan.

Arnold, the head of the air force, had partly similar, partly different motivations. He had a rather unusual place in the JCS structure. As the USAAF was still part of the army, Arnold was actually Marshall's subordinate and was usually careful to support Marshall's positions in larger meetings. However, Marshall was far more interested in tactical airpower over strategic, and was sceptical about the USAAF's claims about the efficacy of the latter.[23] As such, he and allowed Arnold huge latitude in the control of air force operations.[24] Both Leahy and King also were also more than happy to treat Arnold as an equal member of the JCS and rarely questioned him on questions of air strategy.[25] Arnold was the member of the JCS who seemed most motivated by thoughts of revenge against the Japanese. Members of his aircrews were being publicly executed by the Japanese in 1944 and 1945, as the US strategic bombing campaign started to lay waste to Japanese cities. During his trips to the Pacific, he heard stories of these executions and other horrific war crimes committed by the Japanese, reacting emotionally.[26] One would have thought that this would have made Arnold central to the debate over the atomic bomb, as it was clear that the new weapon would have to be delivered by one of Arnold's strategic bombers.

However, Arnold, even though he knew of the Manhattan Project from its beginning, only mentioned the bomb's existence just before it was used and when he did so, remained coy as to his own views. Some of this was due to his weak heart. For significant stretches between 1943 and 1945, Arnold had to take leave after suffering a series of heart attacks. He missed the important White House meeting on 18 June because he felt he could not rush back from a visit he was making to the Pacific theatre. He did not want to put his heart under too much pressure.

Like King, however, he was also conflicted about the atomic bomb because he did not want to detract from what he saw as the USAAF's specific role in ending the war against Japan. Arnold was a great believer in the power of strategic bombing and had specifically put General Curtis LeMay in charge of the air force's campaign against Japan. LeMay, with Arnold's full backing, had controversially turned to the firebombing of

---

[23]John W Huston (ed.), *American Airpower Comes of Age: General Henry H. 'Hap' Arnold's World War II Diaries*, Vol 1 (Maxwell: Air Univ Press 2002), 239–40.

[24]Richard G. Davis, *Carl A. Spaatz and the Air War in Europe* (Washington: Centre for Air Power History 1993). 496.

[25]Phillips Payson O'Brien, *How the War was Won: Air-Sea Power and Allied Victory in World War II* (Cambridge: Cambridge Univ Press 2015). Chapter 4 has a discussion of the different members of the JCS and their interactions.

[26]Henry A. Arnold Mss, Library of Congress, Washington, DC, Reel 3, Diary entry, 16 June 1945.

Japanese cities.[27] Arnold definitely wanted it seen that the air force had already done the major job weakening Japan before either the atom bomb was dropped or an invasion took place.[28]

Unlike King, however, Arnold was involved in the final process of the decision to drop the bomb, which took place during the famous Potsdam Conference in late July 1945. Of course, he was still careful about the words that he used in his diary. On 23 July, he met with Stimson to confer on what he called the 'ultra' bombing effort.[29] The two discussed a wide range of issues including whether the bomb would force the Japanese to surrender, how it would affect Japanese psychology and what kind of effect it would have on the communities around the target. After meeting with Stimson he wrote, enigmatically; 'Some day someone will dissolve the atom, release the atomic forces and harness the resultant terrific power as a destructive explosive. When?'[30]

The next day Arnold received a detailed update on the plans to drop the atom bomb.[31] It contained all the information that he needed, including the targets, the method of delivery (B-29 bombers) and the type and estimated time of the attack (between 1 and 10 August). Four cities were listed as possible targets, with Hiroshima and Nagasaki the highest priorities. Both were described as large industrial cities, with the implication that they had large civilian populations. There was even an interesting side note to the effect that it was thought that a large number of Japanese industrialists and political figures had taken refuge in them as they had heretofore not been heavily attacked.

So, it was certainly clear to Arnold that a large number of civilians would be killed in the attacks. This bothered him little, if at all, at the time, and he was happy to allow his deputy, General Carl Spaatz to be given operational control of the bombs.[32] Spaatz was serving as the head of all US strategic bombing forces, and it was once the weapon was released to his overall control that the clock started ticking towards the attack on Hiroshima. He was actually far more open than Arnold in his doubts about dropping the bomb, and requested a written order before he agreed to implement the plan to cross the atomic threshold.[33]

---

[27]Michael Sherry, *The Rise of American Airpower: The Creation of Armageddon* (New Haven: Yale Univ Press 1987), 266–83.

[28]There is a meditation on Arnold's reasons for missing the 18 June meeting in the edited version of his diary. See Huston (ed.), *American Airpower Comes of Age*, Vol 2, 318–19.

[29]Henry A. Arnold Mss, Library of Congress, Washington, DC, Reel 3, Diary entry 23 July 1945.

[30]Henry A. Arnold Mss, Library of Congress, Washington, DC, Reel 3, Diary entry 23 July 1945.

[31]John Stone to Arnold, 24 July 1945, Accessed online through the Truman Library: https://www.trumanlibrary.org/whistlestop/study_collections/bomb/large/documents/index.php?documentid=31&pagenumber=1.

[32]Henry A. Arnold, *Global Mission* (New York: Harper and Bros 1949), 589. See also: Thomas M. Coffey, *HAP: The Story of the US Air Force and the Man Who Built It, General Henry H. 'Hap' Arnold* (New York: The Viking Press 1982), 382.

[33]Schaffer, *Wings of Judgement*, 147.

Arnold's true feelings about the bomb also become complicated by his immediate and longer term reactions to the attacks on Hiroshima and Nagasaki. When the first confirmed news was received that the attacks were successful, Arnold was positive, agreeing that they were proper payback for the suffering that the Japanese had inflicted on American prisoners of war.[34] However, as time went by, some have argued that Arnold's view on the atom bomb became more negative.[35] Others, however, dispute this.[36]

What seems hard to support is the idea that Arnold had a strong ethical objection to the bomb. When he did speak negatively, it was to stress that Japan was already defeated, the implication being that the bomb was not necessary to force them to surrender. It seems more of a need to show how much the USAAF had already done to drive Japan out of the war then to imply that an ethical wrong had been committed. In Arnold's memoirs, for instance, there is no discussion of him being opposed to the dropping of the atomic bomb on ethical grounds. The most important point that he seems keen to make in his memoirs was that the atom bomb had to be delivered by aircraft.[37]

There might, as in the case of Ernest King, have been a residual doubt in his mind about whether the USA should have crossed the atomic threshold, but this cannot obscure the fact that for the last few weeks in July Arnold was involved in the command decisions about the use of the bomb, and was willing to let the plan go ahead without objection.

Marshall and Leahy: the hard poles

If King and Arnold present somewhat obscure and complex reactions to the idea of atomic warfare in 1945, the two most powerful members of the joint chiefs, Marshall and Leahy, present clearer, contrasting pictures. By this time in the war they were the yin and yang of American grand strategy. Between 1942 and 1945, they had often argued for diametrically different ways of war. Marshall favoured a large-army strategy, based on attacking Germany first with a landing in France as soon as possible and, after that, an invasion of Japan as soon as one was practicable. Leahy, on the other hand, had a more economic and indirect notion of how victory would be won. He wanted the Allies to concentrate on establishing complete air and sea supremacy in both

---

[34]Ed Cray, *General of the Army: George C. Marshall: Soldier and Statesman* (New York: Norton 1990), 548.
[35]Dik Alan Daso, *Hap Arnold and the Evolution of American Air Power* (Washington: Smithsonian Institute Press 2000), 209.
[36]Coffey, *HAP: The Story of the US Air Force*, 371.
[37]Arnold, *Global Mission*, 233. Arnold discusses the atomic bomb on six pages in this memoir. The one mention of any opposition to the explosion of the bomb occurred when one of the scientists came to Arnold trying to have the test explosion stopped for the reason that they could not be sure how powerful the bomb would be (255).

Europe and the Pacific, and then once having choked them off, compel German and Japanese surrender. Within this vision was Leahy's belief that American casualties and attacks on civilian targets should be kept to an absolute minimum.

This meant that Leahy and Marshall fought over not only the atom bomb in the spring and summer of 1945, but also about the need to invade Japan. Marshall's views were those of a direct soldier. Within the Joint Chiefs, he was the most aggressive in pressing for Truman to authorise the plan for the Kyushu attack (codenamed *Olympic*).[38] His arguments for *Olympic* were initially based on somewhat optimistic casualty projections. During the meeting in the White House on 18 June, Marshall opted to present to the president the rosiest possible picture of expected US casualties for *Olympic*, approximately one American casualty for each five Japanese soldiers to be fought.[39] It is important to see Marshall being so optimistic about casualties, because this contradicts one of his most famous statements about dropping the atomic bombs, that it was done to save half a million US casualties.

Because of Marshall's seniority in the War Department, of all the chiefs he played the most prominent official role in the decision on whether and how to use the atom bomb. While not a standing member of the Interim Committee, Marshall was invited to attend during two of its most important meetings, those on 31 May and 1 June. Marshall was invited specifically because Truman had asked that he be involved in making the final decision on the use of the bomb.[40] Marshall had known of the atomic bomb programme from the start, and therefore had already given a good deal of thought to how it could be integrated into his strategic vision. Two days before, he attended the first interim committee meeting, he had a detailed discussion on precisely this question with assistant secretary of war, John McCloy. McCloy recorded Marshall's thoughts, and demonstrated that the general saw the atom bomb as a new weapon, which shared some ethical and strategic similarities with poison gas, but which, crucially, was a legitimate weapon of war to be used if the US government decided it was needed to end the conflict.

---

[38]Truman, *Memoirs: Volume One*, 416.

[39]Frank Settle, *General George C. Marshall and the Atomic Bomb* (Santa Barbara: Praeger 2016), 92. Settle describes Marshall as 'evasive' during the meeting. Marshall claimed that the best comparison for *Olympic* was General MacArthur's campaign on Leyte, where US–Japanese casualties were running at a rate of 1–5. There were far bloodier comparisons, including Okinawa and Iwo Jima, which were much closer to 1–2 or even 1–1.25. 'Minutes of Meeting held at the White House, 18 June 1945', Accessed online through the Truman Library. https://www.trumanlibrary.org/whistlestop/study_collections/bomb/large/documents/index.php?documentid=21&pagenumber=1.

[40]'Notes of the Meeting of the Interim Committee, 31 May 1945', p. 2.

General Marshall said he thought these weapons might first be used against straight military objectives such as a large naval installation and then if no complete result was derived from the effect of that, he thought we ought to designate a number of large manufacturing areas from which the people would be warned to leave – telling the Japanese that we intended to destroy such centers. There would be no individual designations so that the Japs would not know exactly where we were to hit – a number should be named and the hit should follow shortly after. Every effort should be made to keep our record of warning clear. We must offset by such warning methods the opprobrium which might follow from an ill considered employment of such force.

The General then spoke of his stimulation of the new weapons and operations people to the development of new weapons and tactics to cope with the care and last ditch defense tactics of the suicidal Japanese. He sought to avoid the attrition we were suffering from such fanatical but hopeless defense methods – it requires new tactics. He also spoke of gas and the possibility of using it in a limited degree, say on the outlying islands where operations were now going on or were about to take place. He spoke of the type of gas that might be employed. It did not need to be our newest and most potent – just drench them and sicken them so that the fight would be taken out of them – saturate an area, possibly with mustard, and just stand off. He said he had asked the operations people to find out what we could do quickly – where the dumps were and how much time and effort would be required to bring the gas to bear. There would be the matter of public opinion which we had to consider, but that was something which might also be dealt with. The character of the weapon was no less humane than phosphorous and flame throwers and need not be used against dense populations or civilians – merely against these last pockets of resistance which had to be wiped out but had no other military significance'.[41]

It was certainly one of the most important reflections given by a US military leader on the bomb (and weapons of mass destruction as a whole) during the war. It has been used occasionally to say that Marshall had doubts about the use of atomic weapons. However, the most revealing thing was Marshall's surprising willingness to contemplate using weapons such as the atom bomb and gas. During the war, all the major powers debated whether they should use gas attacks, though in the end no one was willing to do so against another major power (there were examples of the Japanese using different gasses in their brutal war against the Chinese).[42] Marshall, however, was the most senior American who seemed most supportive of using gas if it was thought practicable. Two weeks after he spoke to McCloy, Marshall sent

---

[41]'Memorandum of Conversation', 29 May 1945, McCloy Notes. Accessible through Marshall Foundation Online: http://marshallfoundation.org/library/digital-archive/memorandum-of-conversation/.
[42]Yuki Tanaka, 'Poison Gas: The Story Japan would Like to Forget', *Bulletin of Atomic Scientists* 44/8 (1988), 10–19.

King an Army memorandum which called for the pre-emptive use of chemical warfare against the Japanese during *Olympic*.[43]

In fact, it was Marshall who from that moment provided the most consistent support for the use of atomic weapons. During the two meetings of the Interim Committee, which he attended – the ones which determined the direction of weapons usage and during Byrnes claimed that a consensus had been reached that the atom bomb should be used without warning on an industrial target that contained a large number of worker's houses – Marshall seemed completely comfortable with the idea that the bomb would be used.[44] During the 31 May meeting, he spoke almost entirely on the atomic bomb's place in US–USSR relations.[45] During the 1 June meeting, the one in which Byrnes made his comment about their being a strong consensus on using the bomb, Marshall was silent.[46]

In the run up to the use of the bomb Marshall's silent acceptance moved to a far more active assent. By the time of Potsdam, Marshall was eager to use the atom bomb. He had no faith that strategic bombing would end the war on its own, and finally understood that the invasion of Japan would now be far bloodier than he had claimed in June.[47] He admitted (indirectly) to Truman that his earlier assumptions of relatively light losses from an invasion of Japan were wrong, and was now telling the president that it might cost half a million US casualties to force the Japanese to surrender by invading and conquering the home islands.[48] He was so worried about the resistance the Japanese could pose that he started considering using the atomic bomb as a tactical weapon against Japanese troops on Kyushu.[49] Thankfully such a disastrous move never occurred.

After the war, Marshall continued with his strong support for using the atomic bomb. He provided more detail than he ever had previously in some interviews given to his biographer, Forrest Pogue, not long before his death. By this time, Marshall was more than happy to describe the use of the atomic bomb as 'wise'.

Q. 71. Do you feel it was necessary to drop the bomb to shorten the war?

A. I regarded the dropping of the bomb as of great importance and felt that it would end the war possibly better than anything else, which it did, and I think

---

[43]George Marshall Mss, George Marshall Foundation Library, Lexington VA, Marshall to King, 15 June 1945.

[44]Settle, *George C. Marshall*, 78–80.

[45]'Notes of the Meeting of the Interim Committee, 31 May 1945', p. 11.

[46]'Notes of the Meeting of the Interim Committee, 1 June 1945'. Marshall is listed as having attended the meeting, but there is no record of him speaking during its proceedings.

[47]Schaffer, *Wings of Judgement*, 166.

[48]Truman, *Memoirs: Volume One*, 417.

[49]Miscamble, *The Most Controversial Decision*, 82, 116.

that all the claims about the bombings and all afterwards were rather silly. Because we had had these terrific destructions and it hadn't had these effects. I think it was quite necessary to drop the bombs in order to shorten the war.

Q. 72. Do you feel in retrospect that it would have been better to refrain from using it?

A. In retrospect, I feel the same way about it. There were hundreds and hundreds of thousands of American lives involved in these things as well as hundreds of billions of money. They had been perfectly ruthless. We had notified them of the bomb. They didn't choose to believe that. And what they needed was shock action and they got it. I think it was very wise to use it.[50]

If Marshall represented the most clear-cut example of someone who supported the use of the atomic bomb, Leahy represented the most clear-cut example of someone who was opposed. Leahy was Marshall's superior, not only in position as Chairman of the Joint Chiefs, but also in rank, having been promoted to the five-star level one day before the general.[51] Leahy had a remarkable position in the US government during the war. As the president's chief of staff, he was the only policy-making individual who had daily, direct access to the Oval Office. He was also, along with Harry Hopkins, the only individual who had complete access to Roosevelt's and Truman's correspondence with world leaders such as Winston Churchill and Joseph Stalin.[52]

Leahy's role in the atomic bomb decision has been rather poorly understood, partly because Marshall liked to pretend he was more powerful than he really was. After the war, Marshall claimed that Leahy was only let in on the atomic secret late in the day, not long before the successful test of the weapon in July 1945.[53] Nothing could be more untrue. Leahy would have been told about the atomic bomb by Roosevelt, his close personal friend, almost immediately after he became FDR's chief of staff in July 1942. In fact, as time came on Roosevelt saw Leahy as the only person he could fully trust with America's atomic secrets. This was made explicitly clear in September 1944, when Roosevelt invited Churchill to a meeting in Hyde Park NY (the president's family home) to discuss atomic policy. During these top

---

[50]'Interview with General Marshall', Tape 14, 11 February 1957, p. 27; Accessed through Marshall Foundation online. http://marshallfoundation.org/library/wp-content/uploads/sites/16/2014/05/Marshall_Interview_Tape14.pdf.

[51]As a five-star member of the navy Leahy was a Fleet Admiral, a rank he was given on 15 December 1944. Marshall was made the army equivalent, General of the Army, the following day.

[52]William Rigdon, *White House Sailor* (Garden City: Doubleday 1962), 6–9; see also: George Elsey, *An Unplanned Life* (Columbia: Univ Missouri Press 2016), 18–21.

[53]'Interview with General Marshall', Tape 14, 11 February 1957, pp. 26–27; Accessed through Marshall Foundation online. http://marshallfoundation.org/library/wp-content/uploads/sites/16/2014/05/Marshall_Interview_Tape14.pdf.

secret meetings the only people in the room were often Roosevelt, Churchill and Leahy.[54] Marshall was kept far away from Hyde Park during this, and all other meetings during the war.[55]

The problem we have with Leahy was the same with which Marshall was presented. Leahy was discrete and so close to the president that he rarely had to write things down. Before Roosevelt died, Leahy spent a good deal of his time trying to shape FDR's opinion about the bomb – in a negative direction. He complained about its cost, predicted it would not work, and believed it was immoral. We know this because he mentioned his beliefs to one of the few people in the US government he trusted, H. Freeman Matthews. Matthews had worked with Leahy when the latter was the US ambassador to Vichy France in 1941. In 1944–1945, he was head of the State Departments Western European desk before becoming Under Secretary of State.

Matthews was one of the few people with whom Leahy is recorded as discussing the atomic bomb, and it was clear not only did Leahy hope it would not work, he believed it would be a tragedy if it did and the USA then used it against an enemy. He told Matthews before it was tested that 'I don't think this thing is going to work, this bomb. But ... if it does, it's going to have terrible, terrible consequences for the future'.[56] Matthews was struck by the intensity of Leahy's loathing for the atom bomb, writing in his unpublished memoirs that the admiral was convinced that the bomb was a 'terrible thing for the world'.[57]

The problem Leahy had was trying to keep the bomb from being dropped once Roosevelt had died. He was not a close friend of Truman and the new president listened to other advisers, such as Byrnes and Marshall, as much if not more than he did to Leahy. That did not stop the admiral from trying. He was in the room when Truman was given his first detailed briefing about the atomic bomb programme's existence and when he and Truman first discussed the weapon, Leahy was very negative. He told the president that it was '...the biggest fool thing that we have ever done'.[58] Truman, however, was unmoved by Leahy's negativity and the admiral was not made a member of the interim committee. Byrnes did, however, keep him abreast of developments, paying two

---

[54]William D. Leahy Mss, Library of Congress, Washington, DC, Leahy Diary, 17–19 Sept, 1944. See also, William D. Leahy, *I Was There: The Personal Story of the Chief of Staff of Presidents Roosevelt and Truman Based on His Notes and Diaries of the Time* (New York: McGraw-Hill 1950), 264–6.

[55]Marshall admitted in an interview after the war that he expected on a number of occasions that he would be invited to Hyde Park or FDR's winter retreat in Warm Springs Georgia, but that no invitation was ever issued. 'Interview with General Marshall', 11 February 1957, p. 18. Accessible through Marshall Foundation online on: http://www.marshallfoundation.org/library/wp-content/uploads/sites/16/2014/05/Marshall_Interview_Tape14.pdf.

[56]This comes from the H. Freeman Matthews oral history interview on the Truman library website, http://www.trumanlibrary.org/oralhist/matthewh.htm, p. 6. (interview given 6 June 1973).

[57]H. Freeman Matthews Mss, Princeton University Archives, Princeton NJ, unpublished memoirs, p. 604.

[58]Truman, *Memoirs: Volume One*, 10–11.

evening visits to Leahy's Washington, DC townhouse to discuss atomic policy with him.[59]

What also makes Leahy's position unusual was that he strongly opposed dropping the bomb and invading Japan at the same time. This position, which might have seemed contradictory, was based on a consistent ethical stand that he had shown for much of his career. Starting with his first experience of combat, back in the Spanish–American War of 1898, Leahy maintained that the USA should not wage war against civilian targets.[60] During Second World War, he continued this position, though it must be said that there was a certain amount of self-deception in his approach. He believed the USAAF when they claimed that they were doing 'precision' attacks against Axis targets.[61] Only when he attended the Potsdam Conference in July did he see the true devastation that had been inflicted by strategic bombing. Leahy's heartfelt opposition to the use of the atom bomb came from the strength of his ethical convictions, which he admitted overrode his more logical side. As he told a close aide '...deep in the heart, I don't like to see weapons like that developed and used'.[62]

As a sign that Leahy's lobbying might have been playing in Truman's mind, it was only after the president and his chief of staff arrived for the Potsdam Conference that Truman wrote in his diary that he did not want the atom bomb used against women and children.

If seeing the devastation in Germany only reconfirmed Leahy's opposition to the use of the atom bomb, he was left with the basic conundrum that the same ethical standard which had him opposing the bomb also led him to argue against any invasion of Japan. Leahy believed that with Japan completely cut off from the rest of the world by US air and sea power, the war was over and the Japanese government would have to surrender – sooner rather than later. As the war was over in his mind, the idea of invading Japan to possibly hasten its end seemed pointless; further slaughter and the death and maiming of hundreds of thousands on both sides that would make little material difference to the war's outcome. It was why he led the opposition to the invasion of Kyushu on 18 June and continued afterwards to impress on the president the terrible cost any such attack would incur.

In many ways he succeeded too well in this task, and made Truman even more eager to use the bomb when it was shown to be operable. Leahy's overall position would have been politically very difficult for Truman to take (were the president even inclined to do it – which he was not). Truman wanted the war to be over, and if the atom bomb offered him an opportunity to speed up that process, he was darn sure he was going to take it.

---

[59]William D. Leahy Mss, Library of Congress, Washington, DC, Leahy Diary, 20 May 1945, 4 June 1945.
[60]Leahy Diary, Vol I, 33–4.
[61]Leahy, *I Was There*, 395–6.
[62]Bernard L Austin Oral History, p. 65, Naval History and Heritage Command Library, Washington, DC.

When the bomb was dropped, Leahy suffered a severe depression. He remarked to those closest to him that the USA had committed a grave ethical mistake, as war was never to be waged directly against civilian targets.[63] To some who knew him, the regret he felt at the bomb's use was so great it represented the moment when his robust health first started to give way. In his memoirs, written 5 years later, he could not contain his regrets. He stated outright that in using the bomb, the USA 'had adopted an ethical standard common to the barbarians of the Dark ages. I was not taught to make war in that fashion, and wars cannot be won by destroying women and children'.[64]

The atom bomb, the end of the war and the lost victory

Writing in 1949, Arnold decried what he saw as the failure of victory in the Second World War to bring a true peace. 'One thing stands out clearly against the background of my experience: the winning of peace is much more difficult than the winning of even a global war. One look at the condition of the poor old world today, 4 years after the supposed ending of Second World War, almost makes me gasp. Where is our peace?'[65] Arnold was giving voice to a feeling that was shared by most of the joint chiefs (and many others). The war was supposed to lead to a longstanding and solid period of peace, and yet within years relations between the USA and the USSR had deteriorated so dramatically that an even more destructive, nuclear war, seemed a real possibility. Leahy, who played a very important role in determining US policy during the Cold War, was maybe even more depressed. 'Victory' to him meant far more than conquering Japan and Germany. As he surveyed the world, he believed that the USA was and would continue 'paying for this war in many ways long after we, and our children too, had passed away. It may require the better part of a century to bind up the wounds of a world torn in its physical structure by forces which were unleashed first by the aggressor nations and then by us'.[66]

It might be surprising to see the chiefs being so ambiguous about the meaning of victory in the war. They had done their job. Germany and Japan had been completely conquered, their armed forces destroyed, and their ability to threaten the USA nullified. Yet, the war was supposed to be more than this, and by a higher standard it fell short. There were two specific ways this was tied to the atomic bomb decision. The first was the meaning of the bomb in the post-war world, and this was what most preoccupied Leahy. What Leahy was most worried about was that if the USA used the bomb in a

---

[63]Henry A. Adams, *Witness to Power: The Life of Fleet Admiral William D. Leahy* (Annapolis: Naval Institute Press 1985), 299.
[64]Leahy, *I Was There*, 441.
[65]Arnold, *Global Mission*, 265.
[66]Leahy, *I Was There*, 438.

first strike attack, it would make it far more likely that it would suffer similarly in the coming years or decades once other powers also developed atomic weaponry. He wanted the USA to send a message to the world that the atom bomb was such a terrible weapon (in his mind morally equivalent or even worse than poison gas) that it should not be used even if it was being targeted on a country that had no retaliatory capability. After the war, he continued his crusade against atomic weapons. He wrote one of the first detailed definitions of weapons of mass destruction in the US government.[67] Until he left office in early 1949, he did his best to ensure that the USA did not have a formally approved first strike atomic policy.

Then there was the question of the meaning of the bomb, victory and the Cold War. Where the chiefs were different from many in the civilian hierarchy, most prominently Byrnes, was that even those who supported using the bomb had no desire that it be seen as a weapon against the USSR, a way to use 'victory' to intimidate the Soviets. They saw victory as being over Naziism and Imperial Japan. They were less engaged with starting a new competition with the USSR. Though they had some differences in their views of the atom bomb, they had little or no interest in using the bomb to start a competitive new era over the USSR, to manipulate victory as part of an early Cold War. This difference, perhaps more than any other, showed the difference of the military mind from the civilian.

In the end, two very important differences stand out between the military and the civilian leadership. Compared to Truman, Byrnes and Stimson – Leahy, Marshall, King and Arnold had not only a wider range of opinion they had, perhaps most surprisingly, more doubts about the use of the bomb. Three of the four mentioned different concerns about the bomb after the war and one, Leahy, spoke out against it before it was dropped. Their opposition emerged from different impulses; ethical, practical and to a certain degree selfish. Arnold and King, for instance, were both concerned with protecting the reputation of their branch of the armed service in causing the defeat of Japan. They did not want the atom bomb to obscure what they believed were the decisive roles of the USAAF and USN in bringing Japan to defeat. Both also had some ethical doubts about using the weapon, and were certainly uneasy discussing it after the war. However, in both cases, it is important not to overstate the ethical component in their thinking. Leahy on the other hand, had enormous ethical problems with dropping the bomb, and publicly aired them in his memoirs which were published not long after he left office. In fact, Leahy was the person at the upper echelon of the US government most opposed to the use of atomic weapons. No civilian matched the intensity or duration of his opposition.

---

[67]William D. Leahy Mss, Naval History and Heritage Command, Washington, DC, Leahy Memorandum for JCS, May 1947.

Even Marshall, who was the most robust in arguing for the use of the atomic bomb (and that of poison gas) was found to have made qualifying comments a few months before Hiroshima, in which he said the bomb should first be used on a strictly military target and only later be used to attack civilians if the Japanese had still not surrendered.

Why did the military debate differ so from the civilian and what did it have to say about the end of the war? To begin with, the debate showed that the military, or at least some of its leadership at the JCS level, did not see the Second World War as a war which would automatically transform the USA into a permanent global police force. They were much more concerned with the defeat of the Axis Powers, and in protecting what they saw as the reputation of the USA and its armed services for fighting wars properly and effectively in doing so. They had little interest in making a grand global statement about American power, to intimidate, for instance, the USSR. In other words, again it was the military leadership of the USA that was least likely to see the world war mutating quickly into a Cold War. In this way, the Joint Chiefs present a very different debate from that which has obsessed many historians for the past decades. Intimidating the Soviet Union had only miniscule impact on their thinking. They rarely, if at all, connected the bomb to an attempt to limit Soviet expansion. Only George Marshall ever addressed the question before the attacks on Hiroshima and Nagasaki, and even he did it in a non-committal way. Certainly compared to Byrnes or even Truman, the military leadership of the USA was more restrained and less aggressive when it came to planning for the post-war world.

It leaves us with one of the great questions about the end of the war. Had the US military had its way, the war might have ended very differently. The bomb might not have been dropped and the war could have continued on for months longer until either an invasion of Kyushu, the entry of the USSR into the conflict or even an extended air-sea blockade would have forced the Japanese to surrender. The atomic threshold would not have been crossed, but on the other hand, far more Japanese, both military and civilian, would have died had an invasion or extended blockade occurred. There is no easy answer in this debate, as there was no easy answer at the time.

Disclosure statement

No potential conflict of interest was reported by the author.

Bibliography

Adams, Henry A., *Witness to Power: The Life of Fleet Admiral William D. Leahy* (Annapolis: Naval Institute Press 1985).

Alperovitz, Gar, *Atomic Diplomacy: Hiroshima and Potsdam, the Use of the Atomic Bomb and American Confrontation with Soviet Power* (New York: Vintage 1965).

Alperovitz, Gar et al., *The Decision to Use the Atomic Bomb: And the Architecture of an American Myth* (New York: Alfred Knopf 1995).

Arnold, Henry A., *Global Mission* (New York: Harper and Bros 1949).

Berstein, Barton, 'A Postwar Myth: 500,000 US Lives Saved', *Bulletin of the Atomic Scientists* 42 (June/July 1986), 38-40.

Bird, Kai and Lawrence Lifschultz (eds.), *Hiroshima's Shadow* (Stony Creek: The Pamphleteers Press 1998).

Buell, Thomas B., *Master of Sea Power: A Biography of Fleet Admiral Ernest J. King* (Annapolis: Naval Institute Press 1980).

Coffey, Thomas M., *HAP: The Story of the US Air Force and the Man Who Built It, General Henry H. "Hap" Arnold* (New York: The Viking Press 1982).

Costigliola, Frank and Michael J. Hogan (eds.), *America in the World: The Historiography of American Foreign Relations since 1941* (Cambridge: Cambridge Univ Press 2014).

Craig, Campbell and Sergey Radchenko, *The Atomic Bomb and the Origins of the Cold War* (New Haven: Yale Univ Press 2008).

Cray, Ed, *General of the Army: George C. Marshall: Soldier and Statesman* (New York: Norton 1990).

Daso, Dik Alan, *Hap Arnold and the Evolution of American Air Power* (Washington: Smithsonian Institute Press 2000).

Davis, Richard G., *Carl A. Spaatz and the Air War in Europe* (Washington: Centre for Air Power History 1993).

Elsey, George, *An Unplanned Life* (Columbia: Univ Missouri Press 2016).

Frank, Richard B., *Downfall: The End of the Imperial Japanese Empire* (New York: Penguin 2001).

Giangreco, D. M., *Hell to Pay: Operation Downfall and the Invasion of Japan 1945-1947* (Annapolis: Naval Institute Press 2009).

Huston, John W (ed.), *American Airpower Comes of Age: General Henry H. "Hap" Arnold's World War II Diaries*, Vols. 1–2 (Maxwell: Air University Press 2002).

King, Ernest J. and Walter Muir Mitchell, *Fleet Admiral King: A Naval Record* (London: Eyre and Spottiswoode 1953).

Leahy, William D., *I Was There: The Personal Story of the Chief of Staff of Presidents Roosevelt and Truman Based on His Notes and Diaries of the Time* (New York: McGraw-Hill 1950).

Maddox, Robert James, *Weapons for Victory: The Hiroshima Decision* (Columbia: Univ Missouri Press 2004).

Malloy, Sean L., *Atomic Tragedy: Henry L. Stimson and the Decision to Use the Atomic Bomb against Japan* (Ithaca: Cornell Univ Press 2018).

Miscamble, Wilson, *The Most Controversial Decision: Truman, the Atomic Bombs and the Defeat of Japan* (Cambridge: Cambridge Univ Press 2011).

Morison, Elting E., *Turmoil and Tradition: A Study of the Life and Times of Henry L. Stimson* (Boston: Houghton Mifflin 1960).

O'Brien, Phillips Payson, *How the War Was Won: Air-Sea Power and Allied Victory in World War II* (Cambridge: Cambridge Univ Press 2015).

Offner, Arnold A., *Another Such Victory: President Truman and the Cold War 1945-1953* (Palo Alto,: Stanford Univ Press 2002).

Rigdon, William, *White House Sailor* (Garden City: Doubleday 1962).

Samuel Walker, J., *Prompt and Utter Destruction: Truman and the Use of Atomic Bombs against Japan* (Chapel Hill: UNC Press 1997).

Schaffer, Ronald, *Wings of Judgement: American Bombing in World War II* (New York: Oxford Univ Press 1985).

Settle, Frank, *General George C. Marshall and the Atomic Bomb* (Santa Barbara: Praeger 2016).

Sherry, Michael, *The Rise of American Airpower: The Creation of Armageddon* (New Haven: Yale Univ Press 1987).

Skates, John R., *The Invasion of Japan, Alternative to the Bomb* (Columbia: Univ South Carolina Press 1994).

Stoler, Mark A, *Allies and Adversaries, the Joint Chiefs of Staff, the Grand Alliance, and US Strategy in World War II* (Chapel Hill: UNC Press 2000).

Takaki, Ronald, *Hiroshima: Why America Dropped the Bomb* (Boston: Little Brown 1995).

Tanaka, Yuki, 'Poison Gas: The Story Japan would Like to Forget', *Bulletin of Atomic Scientists* 44/8 (1988), 10–19.

Truman, Harry S., *Memoirs: Volume One, Year of Decisions* (Garden City: Doubleday 1955).

Werrell, Kenneth P., *Blankets of Fire: US Bombers over Japan during World War II* (Washington: Smithsonian Institute Press 1996).

Winters, Francis X., *Remembering Hiroshima: Was It Just?* (Farnham: Ashgate 2009).

Slow failure: Understanding America's quagmire in Afghanistan

Christopher D. Kolenda

ABSTRACT

The United States government has no organised way of thinking about war termination other than seeking decisive military victory. This implicit assumption is inducing three major errors. First, the United States tends to select military-centric strategies that have low probabilities of success. Second, the United States is slow to modify losing or ineffective strategies due to cognitive obstacles, internal frictions, and patron-client challenges with the host nation government. Finally, as the U.S. government tires of the war and elects to withdraw, bargaining asymmetries prevent successful transitions (building the host nation to win on its own) or negotiations.

After a promising start – the rapid overthrow of the Taliban regime – the war in Afghanistan has dragged on for over 18 years with no end in sight. In fact, a September 2018 U.S. Department of Defence assessment reports that the Taliban controls or contests nearly 50 percent of the country. Why has the U.S. intervention in Afghanistan become a quagmire?

An important part of the answer is that the United States government has no organised way of thinking about war termination other than seeking decisive military victory. This implicit assumption is inducing three major policy and strategy errors that prevent the world's most powerful country from succeeding against poorly trained militants.[1] First, the United States tends to select military-centric strategies that have low probabilities of success. Second, the United States is slow to modify losing or ineffective strategies due to cognitive obstacles, internal frictions, and patron-client challenges with the host nation government. Finally, as the U.S. government tires of the war and elects to withdraw, bargaining asymmetries prevent

---

[1]The major interventions in Afghanistan, Vietnam and Iraq reveal similar errors. In all three conflicts, the United States withdrew without securing a durable peace, and handed over an unstable situation to a deeply compromised host-nation partner. Christopher D. Kolenda, 'End-Game: Why American Interventions Become Quagmires, Doctoral Thesis (King's College London 2017).

successful transitions (building the host nation to win on its own) or negotiations. This article will show how each of these three problems became manifest in the war in Afghanistan.

What is war termination?

War termination is the method by which a combatant ends participation in organised hostilities. This can be accomplished in the surrender of one side, a negotiated outcome that creates a mixed or variable sum result, or when a combatant ends participation unilaterally while the conflict goes on. A war termination outcome that results in the durable achievement of a combatant's aims, by definition, means the war was successful.

Decisive victory seems to be the clearest path to success. However, none other than Prussian military theorist Carl von Clausewitz recognised that decisive victory is not always possible or desirable. 'It is possible to increase the likelihood of success without defeating the enemy's forces,' he argues, 'many roads lead to success [and] they do not all involve the opponent's outright defeat.'[2] This statement suggests that the likelihood of success in certain wars can be *increased* by seeking outcomes other than decisive victory. Conversely, fixating on decisive victory might *reduce* the possibility of success in those conflicts. This insight helps to explain America's problems during major interventions against insurgencies.

The U.S. government, however, has no authoritative doctrine that outlines war-termination options beyond decisive victory. What alternatives exist? In broad terms, successful outcomes for interventions against insurgencies may include:

- **Decisive victory**, in which the insurgency capitulates or ceases to exist. This is a zero-sum outcome.
- **Negotiated settlement** or mixed outcome, in which neither side vanquishes the other or forces their surrender. Parties compromise to end the war and settle remaining differences through peaceful politics. Success occurs if the negotiated settlement enables the combatant to achieve its main war aims. This can be a variable sum outcome.
- **Transition**, in which the intervening power degrades the insurgency while building the capacity of the host nation government and security forces. As these forces become superior to the enemy (the so-called crossover point), the intervening power hands over security responsibility to the host nation and withdraws without concluding a peace agreement.[3]

---

[2]Carl von Clausewitz, *On War*, edited and translated by Michael Howard and Peter Paret (Princeton, NJ: PUP 1984), 92, 95, 99.
[3]Kolenda, 'End-Game,' 21–22.

The decision on which war-termination method is the most realistic path to success should inform the strategy. A strategy to achieve a successful transition, for instance, will differ from one oriented on decisive victory. The former would arguably place the highest priority on host government and security force legitimacy, while the latter would prioritise military actions aimed at defeating the enemy's forces. A strategy seeking a negotiated outcome should prioritise diplomatic efforts. Getting the strategy wrong risks prolonging the war and its costs.

Existing empirical analysis could reduce the risk of decision-making errors on which outcome to seek. For interventions against an insurgency, two questions are critical: is the insurgency sustainable and is the host nation government able to win the battle of legitimacy? In their analysis of 71 insurgencies since 1944, RAND scholars Chris Paul et al examine cases for common themes, approaches, and practices that lead towards success for the counterinsurgent. They classify 42 as insurgent wins (a 59% success rate), and tally 29 for the counterinsurgent (a 41% success rate).[4] Each counterinsurgent win required both tangible support reduction of the insurgency and sufficient host nation commitment and motivation to winning the battle of legitimacy. Absence of one factor or the other consistently led to a counterinsurgent loss.[5] For decisive victory to be possible, the insurgency must be cut off from sanctuary, and the host nation government must win the support of the local populace in insurgent controlled and contested areas. For an intervention that envisions replacing one regime with another, prevention of sanctuary and of behaviours that undermine government legitimacy are essential for success.

A myopic strategy – Winning the battle but not the war

The U.S. government believed they had achieved a decisive victory when the Taliban regime was overthrown and a new government installed in its place. At that point, the Bush administration focused on the hunt for al Qaeda and Taliban senior leaders, while the international community rebuilt Afghanistan. The Bush administration was keen to re-orient attention to the next phase in the Global War on Terrorism: the war in Iraq.[6]

The Administration, however, did not adequately consider war termination options in its policy and strategy. For the regime change to be a durable success, the Administration needed to prevent a sustainable insurgency from developing and the new government (and international forces) from

---

[4]Christopher Paul, Colin P. Clarke, Beth Grill, and Molly Dunigan, *Paths to Victory: Lessons from Modern Insurgencies* (Washington DC: RAND 2013), 18.
[5]Paul et al., *Paths to Victory*, 149.
[6]Vanda Felbab-Brown, 'Slip-Sliding on a Yellow Brick Road: Stabilization Efforts in Afghanistan', *Stability: International Journal of Security and Development* 1/1 (2012), 4–19.

undertaking actions that undermined its own legitimacy. Understanding the nature of the Taliban, its relationship with al Qaeda, and the perspectives of Afghanistan's neighbours were critical to preventing an insurgency. Understanding the nature of the Islamic State of Afghanistan (ISA) and the Northern Alliance was essential in preventing the new government from sabotaging itself in predatory behaviour and score-settling.

Problems at the outset

The ISA was a collection of former *mujahideen* leaders that had overthrown the Afghan communist regime in 1992 and was itself overthrown by the Taliban in 1996. It consisted of so-called warlords who often fought among themselves for power and many became notorious for their gross abuses against the Afghan people. Getting rid of the warlords was the most popular achievement of the Taliban. Once in power, the Taliban were as bad or worse than their predecessors. Several factions from the ousted ISA formed the Northern Alliance and continued fighting the Taliban regime.

The United States partnered with the Northern Alliance in 2001 to oust the Taliban. The Bush administration, however, did not consider the risk that super-empowering the so-called warlords could prolong the war and damage U.S. interests.

The Bonn Process was established by the United Nations to form a new government in Afghanistan. The first step was the Emergency Loya Jirga, which was to form an Afghanistan Transitional Administration (ATA) that would write a new constitution and hold presidential elections in 2004 and parliamentary elections in 2005 to establish a permanent and democratic government. While the international community set about establishing milestones and allocating tasks to donors, the warlords vied for control of the government and to freeze out potential rivals. U.S. Envoy Zalmay Khalilzad reportedly manoeuvred for Hamid Karzai (a southern Pashtun) to lead the ATA.[7] Northern Alliance leaders came to support the proposal, in part because Karzai had no militia or independent base of power. An initiative to invite the Taliban to the ATA was floated and quickly rejected at Northern Alliance urging.[8]

The warlords' power and influence during the Emergency Loya Jirga was evident as they gained control of all but two ministries in the ATA. Feeling marginalised during the proceedings, many southern Pashtuns (the traditional Taliban constituency) walked out.[9] Their significant wealth, influence with the

---

[7]Thomas Ruttig, 'The Battle for Afghanistan: Negotiations with the Taliban: History and Prospects for the Future', *New America Foundation* (2011), 3.

[8]Anders Fänge, 'The Emergency Loya Jirga', in Martine van Bijlert and Sari Kouvo (eds.), *Snapshots of an Intervention: The Unlearned Lessons of Afghanistan's Decade of Assistance (2001–2011)* (Afghan Analysts Network 2012), 2–4.

[9]Fänge, 'The Emergency Loya Jirga', 13–17.

international community, and well-armed militias enabled the warlords to ensure their representatives were present, to buy votes as necessary, and to intimidate opposition.[10] To many Afghans, the despised ISA had returned.

The warlords also angled to block interim Afghan Transitional Administration President Hamid Karzai's efforts towards peace. Karzai's representatives had negotiated with Taliban representatives in December 2001 what amounted to be surrender by the latter. Rumsfeld flatly and publicly rejected it during a news conference on 7 December.[11] 'There are a lot of fanatical people,' he concluded, 'And we need to finish the job.'[12] Secretary of State Colin Powell's suggestion that the U.S. seek talks with moderate Taliban was ridiculed as naïve by Foreign Minister Abdullah Abdullah and ultimately rejected by the Bush Administration.[13] The Taliban sent several peace overtures in the early years,[14] but negotiations were rejected by the U.S. and the Northern Alliance factions in the Afghan government.[15] Brahimi later described this decision to be a fundamental error in the Bonn process.[16] Taliban leaders who turned themselves in were sent to prisons in Bagram and Guantanamo, some reportedly tortured.

Adding to the challenges, neither the United States nor the nascent Afghan government coordinated a strategy for a durable political outcome. This eventually lead to dangerously misaligned interests and incentives. The United States wanted to leave as quickly as possible, while Afghan officials and warlords sought to keep U.S. troops and money flowing into Afghanistan. The Bush administration, moreover, was divided on the extent to which the U.S. should remain involved. As early as November 2001, the State Department was calling for international efforts to rebuild Afghanistan, but the Bush administration remained divided on the matter.[17] It hoped that accelerating the development of Afghan security forces would enable a faster drawdown of U.S. forces.[18]

---

[10]Thomas Ruttig, 'Flash to the Past: Power play before the 2002 Emergency Loya Jirga', (Afghan Analysts Network, 27 April 2012).

[11]Brian Knowlton, 'Rumsfeld Rejects Plan To Allow Mullah Omar "To Live in Dignity": Taliban Fighters Agree to Surrender Kandahar', The New York Times, 7 December 2001, https://www.nytimes.com/2001/12/07/news/rumsfeld-rejects-planto-allow-mullah-omar-to-live-in-dignity-taliban.html.

[12]Thom Shanker, 'Rumsfeld Pays Call on Troops and Afghans', New York Times, 17 December 2001, https://www.nytimes.com/2001/12/17/world/a-nation-challenged-the-secretary-rumsfeld-pays-call-on-troops-and-afghans.html.

[13]Pamela Constable, 'U.S. Hopes to Attract Moderates in Taliban', The Washington Post (17 October 2001), https://www.highbeam.com/doc/1P2-476553.html.

[14]James Dobbins and Carter Malkasian, 'Time to Negotiate in Afghanistan: How to Talk to the Taliban', Foreign Affairs 94/4 (July–August 2015), 53–64.

[15]Barnett R. Rubin, 'Saving Afghanistan', Foreign Affairs 86/1 (Jan/Feb 2007), 58.

[16]Lakhdar Brahimi, 'State Building in Crisis and Post-Conflict Countries', 7th Global Forum on Reinventing Government Building Trust in Government, Vienna, Austria, 26–29 June 2007, 13.

[17]Condoleezza Rice, No Higher Honor: A Memoir of My Years in Washington (New York: Random House 2011), 96, 148; Donald H. Rumsfeld, Known and Unknown: A Memoir (New York: Sentinel 2011), 398; Woodward, State of Denial, 192–5, 220, 237.

[18]The White House, 'Joint Statement on New Partnership Between U.S. and Afghanistan' (28 January 2002).

International and U.S. bureaucratic silos

In an effort to free up the U.S. military for potential use elsewhere and to gain international resources and support for Afghanistan, the Bush administration supported an April 2002 Group of Eight (G8) initiative known as the 'lead nation concept' for the Afghan security sector. This entailed five interdependent lines of effort. Each one was assigned to a donor nation: Afghan National Army (U.S.), Afghan National Police (Germany), Counter-narcotics (U.K.), judiciary (Italy), DDR (Japan).[19] This initiative distributed the requirements to different countries but failed to establish a national or international command authority that could integrate and manage the full range of these efforts. Instead, lead-nation silos developed that were quickly and ably exploited by Afghan elites to promote personal and political power.[20]

For instance, international officials were often blind to the scrimmage for power that dominated efforts to control the security forces. Each ministry developed a strong, informal chain of command tied to the warlords who exercised influence over key appointments.[21] After promoting himself to Marshal, Northern Alliance leader Mohammad Fahim became Minister of Defence and engineered recruiting and training efforts to lock in his control of the Army and freeze out any rivals. Of the first 100 general officers, 90 were from his native Panjshir province.[22]

The Ministry of Interior, meanwhile, was responsible for the police, subnational governance, and counter-narcotics. German and U.S. officials lacked a common view and began to pull police development in different directions – the former towards basic western-style law enforcement, the latter towards a paramilitary role.[23] To add to the problems, Afghan power-brokers allocated police chief positions to local strongmen or to the highest bidder. Having paid large sums of money for their positions, these officials were given license to take actions to recoup their investment and turn a personal profit. This was often achieved through misappropriation of customs revenues, aid dollars, or even extortion such as land-theft and kidnapping for ransom.[24] Because U.S. and international military officials worked closely with Afghan military and police, Afghans perceived that these predatory actions were backed by the U.S. military.

---

[19]Barnett R. Rubin, *Afghanistan from the Cold War through the War on Terror* (New York: Oxford University Press 2013), 311–2.

[20]Kolenda, 'End-Game', 188–96.

[21]Mark Sedra, 'Police Reform in Afghanistan: An Overview', Paper presented at the Bonn e-conference on Afghanistan: 'Assessing the Progress of Security Sector Reform', 4–11 June 2003, 34.

[22]Antonio Giustozzi, *The Army of Afghanistan: A Political History of a Fragile Institution* (London: Hurst 2016), 125–132; Paul O'Brien and Paul Baker, 'Old Questions Needing New Answers: A Fresh Look at Security Needs in Afghanistan', Paper presented at the Bonn e-conference on Afghanistan: 'Assessing the Progress of Security Sector Reform', 4–11 June 2003.

[23]Perito, Robert M, 'Afghanistan's Police: The Weak Link in Security Sector Reform', United States Institute of Peace, August (2009).

[24]Sarah Chayes, *Thieves of State: Why Corruption Threatens Global Security* (New York: W.W. Norton 2015), 20–38.

Local elites, meanwhile, duped international forces and intelligence efforts into targeting personal and political rivals.[25] When such manipulation resulted in military operations, affected local leaders and populace grew to believe that the government and international forces were arrayed against them. Reports of torture in American prisons such as Bagram and Guantanamo added to perceptions of hostile foreign occupation among many Afghans. When directed against community leaders, civilian harm had disproportionately large effects in driving the people away from the government and often into the embrace of armed opposition groups.[26] People began to turn against the new government.

Kleptocracy is born

The perception, therefore, that post-Taliban Afghanistan was peaceful was an illusion. The very officials who were responsible for providing security and reporting incidents often perpetrated the violence themselves.[27] The insecurity and civilian harm undermined the legitimacy of the government in the eyes of affected Afghans and enabled insurgent groups to gain critically needed local support. 'The Taliban insurgency started as a grassroots movement in reaction to the repression unleashed by Afghan security forces, private militias and Enduring Freedom units in 2002–3,' political scientist Antonio Giustozzi wrote in a retrospective assessment to NATO. 'Without that repression, it is unlikely that the few Taliban leaders who wanted to fight on would have been able to re-engage.'[28] In the face of U.S. resistance, U.N. envoy to Afghanistan Lakhdar Brahimi continued to push in 2002 and 2003 for greater international military presence to protect Afghan civilians from predatory militias.

Perceiving such actions as a threat to his government's legitimacy, Karzai repeatedly asked for U.S. assistance in confronting various warlords. Since the warlords controlled the nascent army and police, Karzai had no reliable military force he could use to confront the former. The United States, however, refused to support him in a 2002 effort to take on a threat from eastern warlord Pacha Khan Zadran.[29] A year later, Rumsfeld was adamant to CENTCOM Commander John Abizaid, Zalmay Khalilzad and others, 'We do not want him making moves [against warlords] under the mistaken belief that we are going to back him up militarily.'[30]

---

[25]Anand Gopal, *No Good Men Among the Living: America, the Taliban, and the War through Afghan Eyes* (New York: Metropolitan Books 2014).

[26]For an in-depth study on how civilian harm affected the war see Christopher D. Kolenda, Rachel Reid and Christopher Rogers, *The Strategic Costs of Civilian Harm: Applying Lessons from Afghanistan to Current and Future Conflicts* (Open Society Foundations June 2016), 17–28.

[27]Human Rights Watch, *Violent Response: The U.S. Army in al-Falluja* (16 June 2003).

[28]Antonio Giustozzi, 'The Changing Nature of the Insurgency," in *ISAF Strategic Assessment Capability Final Workshop, 10–12 December 2014*, Ver. 0.1, January 2015, 31.

[29]Khalilzad, Zalmay, *The Envoy: From Kabul to the White House, My Journey Through a Turbulent World* (New York: St. Martin's Press 2016 kindle edition), 2618 of 7203.

[30]The *Rumsfeld* Papers, 'Rumsfeld to Abizaid, "Karzai's Strategy on Warlordism"' (15 September 2003).

Nonetheless, after winning the election in 2004, Karzai removed several warlords and their lieutenants from ministerial positions, to include replacing Fahim at Ministry of Defence.[31] The warlords countered. In May 2005 protests erupted across Afghanistan, instigated by a *Newsweek* article alleging that interrogators at U.S. military prison at Guantanamo had desecrated the Koran.[32] Using local muscle to gain electoral advantage, the September 2005 parliamentary elections became a major victory for warlords and strongmen.[33] With this parliamentary power, they could use legislative action to block efforts aimed at undermining their influence, and could extract enormous bribes during confirmation hearings for ministers. This put Karzai in a difficult governing position. He had to negotiate and balance his obligations to the efforts of the international community with the interests of the warlords. Karzai, in effect, bought them off, enabling the warlords to use government appointments and authority to get rich. The kleptocracy was born. By 2006, warlords began to re-emerge in key ministerial positions.

The Taliban strikes back

Meanwhile, Pakistan provided sanctuary to the Taliban to foment an insurgency.[34] Prolific military businesses and logistics companies in Pakistan and the secretive intelligence service, ISI, gave the Taliban ready access to supplies, logistics, and expertise. Vast Afghan refugee camps offered the potential for substantial recruits. Predatory activities by the government and civilian harm by Afghan and coalition forces provided the opportunity to attract the disaffected to the Taliban cause.

The Taliban made another peace overture in 2004, claimed a former Taliban official who was part of the delegation.[35] The Bush Administration still refused. Standing alongside President Karzai in February 2004, Rumsfeld said, 'I've not seen any indication that the Taliban pose any military threat to the security of Afghanistan.' Karzai, noting that he was being contacted daily by Taliban leaders seeking to be allowed to return home, surmised, 'The Taliban doesn't exist anymore. They're defeated. They're gone.'[36] The 2004

---

[31]U.S. Embassy Brussels Cable, 'Coordinator for Afghanistan Quinn Meetings with European Commission, Council' (21 January 2005); Kenneth Katzman, 'Afghanistan: Post-War Governance, Security, and U.S. Policy', Congressional Research Service (28 December 2004), 14–16.

[32]Carlotta Gall, 'Protests Against U.S. Spread Across Afghanistan', *The New York Times*, 13 May 2005, https://www.nytimes.com/2005/05/13/world/asia/protests-against-us-spread-across-afghanistan.html; Sonali Kolhatkar and James Ingalls, *Bleeding Afghanistan: Washington, Warlords, and the Propaganda of Silence* (New York: Seven Stories 2006), 117–68.

[33]Andrew Wilder, 'A House Divided? Analysing the 2005 Afghan Elections' (Afghan Research and Evaluation Unit, December 2005); Interview with Barnett R. Rubin.

[34]Barnett Rubin and Ahmed Rashid, 'From Great Game to Grand Bargain: Ending Chaos in Afghanistan and Pakistan', *Foreign Affairs* 87/6 (November/December 2008), 37–38.

[35]Interviewee Y.

[36]Liz Sly, 'Rumsfeld, Karzai Declare Taliban no Longer a Threat', *Baltimore Sun* (27 February 2004), http://articles.baltimoresun.com/2004-02-27/news/0402270304_1_taliban-kabul-afghanistan.

overture would be the Taliban's last for many years. By 2006, an insurgency that had durable internal and external support was fighting against a predatory, kleptocratic host nation government that was losing legitimacy. Unless the U.S. and Afghan governments could modify their strategy to reverse these problems, the stage was set for an intractable conflict.

Problems modifying an ineffective strategy

The United States was slow to recognise these festering problems and thus had difficulty modifying its strategy. This section examines three key reasons why: cognitive bias, internal political and bureaucratic frictions, and patron-client problems with the Afghan government. Confirmation bias, for instance, is the tendency to ascribe high credibility to evidence that confirms pre-existing theories or beliefs and low credibility to disconfirming data. This can cause decision-makers to miss clear warning signs that serious problems are arising. The tendency for U.S. agencies deployed to Afghanistan to oper-ate in bureaucratic silos reinforced confirmation bias. The silo problem also created major seams and fault lines that were exploited by Afghan officials and power brokers and led to American efforts that could self-synchronise in damaging ways. As tensions rose with the Karzai government over issues such as civilian casualties, Pakistan, and corruption, the aims and incentives of the U.S. and Afghanistan grew dangerously misaligned.

Cognitive bias and bureaucratic silos

That the U.S. military and civilian officials continued to cite examples of progress even as the security situation deteriorated suggests that they suffered from varying degrees of cognitive bias.[37] U.S. and other international civilian and military officials assumed that as political, security, and economic mile-stones were achieved over time, the new Afghan government would become increasingly legitimate and capable to govern and secure itself. As each mile-stone was achieved, and largely on time, the U.S. and international partners believed they were making progress towards a favourable and durable out-come. In-silo metrics were simply added up with others to produce a picture of relentless success. Rises in Taliban violence tended to be dismissed as the last throes of freedom-hating terrorists.

The in-silo progress was unmistakable, but these indicators were not adding up to strategic success. As outlined in the previous section, each political, military, and economic milestone was capably manipulated by power brokers

---

[37]For example, U.S. Department of State, 'President Bush Discusses Progress in Afghanistan, Global War on Terror', (15 February 2007), https://2001-2009.state.gov/p/sca/rls/rm/2007/80548.htm; NATO, *Progress in Afghanistan, Bucharest Summit* (2–4 April 2008).

and elites to ensure the outcome met their interests. As U.S. officials in Kabul and Washington, D.C. stayed in their bureaucratic lanes, no one was comprehending the big picture or seeing the problems that were emerging in the silos' seams and fault lines.

Despite mounting indicators of insecurity, U.S. civilian and military officials in Kabul maintained a minimalist approach to the war in Afghanistan. Retired General Barry R. McCaffrey reported to Rumsfeld in June 2006 that the Afghan security forces were chronically and 'miserably' under-resourced.[38] Fewer than half of the required advisor positions were filled.[39]

Meanwhile, the Taliban had become capable of mounting large-scale attacks. The significant uptick in 2006 violence alarmed American officials. President Bush remarked during a 2007 speech, 'Today, five short years later, the Taliban have been driven from power, al Qaeda has been driven from its camps, and Afghanistan is free. That's why I say we have made remarkable progress.'[40] Nonetheless, he noted the significant Taliban offensives, and decided to increase support to Afghanistan even as he was surging military forces in Iraq. He outlined five major capacity-building efforts to support the Afghan government. He also pledged to work with Pakistani President Musharraf to defeat terrorists and extremists in Pakistan. The Bush Administration misdiagnosed the problems in Afghanistan. Officials focused on inadequate capacity rather than the more fundamental problems with Afghan government legitimacy and Taliban sanctuaries in Pakistan. Strategy in Afghanistan remained largely unchanged.

Surge and withdraw

President Obama, however, was determined to reform it. He undertook a strategy review upon assuming the presidency and announced major changes in a 1 December 2009 speech at West Point. The U.S. government aimed to achieve a secure, stable, sovereign Afghanistan that could defend itself and prevent the re-emergence of terrorist safe havens.[41] He committed additional military and civilian resources under a revised civil-military campaign, while directing diplomatic efforts to urge greater support from Pakistan and to assist in reconciliation between the Afghan government and those Taliban who wanted peace.

The new strategy, however, sent mixed messages which reflected conflicting views within the Administration. The implicit theory of the case was

---

[38]Barry R. McCaffrey, *Academic Report: Trip to Afghanistan and Pakistan* (United States Military Academy, Department of Social Sciences 3 June 2006).

[39]SIGAR, *Reconstructing the Afghan National Defense and Security Forces: Lessons from the U.S. Experience in Afghanistan* (2017), 38–68.

[40]U.S. Department of State, 'President Bush Discusses Progress in Afghanistan'.

[41]The White House, 'Remarks by the President in Address to the Nation on the Way Forward in Afghanistan and Pakistan', Press Release (1 December 2009), https://obamawhitehouse.archives.gov/the-press-office/remarks-president-address-nation-way-forward-afghanistan-and-pakistan.

that building Afghan government and security force capacity, reversing the Taliban's momentum, and clearing them from key areas would enable the Afghan government to move in and win the battle for legitimacy among the population. By building a strategic partnership with Pakistan, the U.S. hoped to change Pakistan's strategic calculus and induce them to reduce and eliminate insurgent sanctuaries. Reintegration and reconciliation efforts were to provide local fighters and Taliban senior leaders a way out of the conflict. These efforts would enable the U.S. to withdraw surge forces beginning in July 2011, end the combat mission by December 2014, and hand-off a residual insurgency to the ANSF.[42] The strategy, in short, rested on three implicit assumptions: 1) the counterinsurgency model used in Iraq could be applied successfully to Afghanistan, 2) the Afghan government could reform itself sufficiently to win the battle of legitimacy in contested and Taliban-controlled areas, and 3) Pakistan would pressure the Afghan Taliban to give up the insurgency. None turned out to be true.

There were good reasons to doubt that the formula in Iraq could be replicated successfully in Afghanistan.[43] The prevailing interpretation was that the surge in forces plus application of the new counterinsurgency doctrine and a reconciliation programme convinced the Sunni tribes to turn against the deeply unpopular al Qaeda in Iraq.[44] No other cases, however, were used for comparison during the Afghanistan review.[45] The sharp downturn in violence in Iraq was more complicated than portrayed by the military at the time and relied on political and social factors that were not present in Afghanistan.[46] Unlike al Qaeda in Iraq, the Taliban emphasised governance and were using the full range of coercion and persuasion to gain control and garner public support.[47]

Patron-client frictions

Severe patron-client problems compounded decision-making frictions and undermined capacity-building efforts. The U.S. and Afghan governments never developed a common strategy for the war. While the United States wanted to win quickly and leave, the new Afghan government focused on

---

[42]For a discussion of the crossover point see Octavian Manea and John A. Nagl, 'COIN is not Dead: An Interview with John Nagl', *Small Wars Journal* (6 February 2012).

[43]For skeptical views in the ISAF staff see Rajiv Chandrasekaran, *Little America: The War Within the War for Afghanistan* (New York: Vintage 2012), 245–6.

[44]Fred W. Baker III, 'Petraeus Parallels Iraq, Afghanistan Strategies', *American Forces Press Services* (28 April 2009), http://archive.defense.gov/news/newsarticle.aspx?id=54107.

[45]Interviewees H, J, L, M, N, P, X.

[46]Thomas Barfield, *Afghanistan*, 277, 285, 337, 339–342; Christopher D. Kolenda, 'Winning Afghanistan at the Community Level', *Joint Force Quarterly*, Issue 56, (1st Quarter 2010), 25–31.

[47]Stanley A. McChrystal, *COMISAF's Initial Assessment* (30 August 2009), 2–6 to 2–8; Taliban, 'Code of Conduct (Layha)', Taliban 2009 Rules and Regulations Booklet (seized by Coalition Forces on 15 July 2009 IVO Sangin Valley 2009).

consolidating power and extending international presence and financial support. Afghanistan remained at the very top of the world's most corrupt governments.[48] Karzai grew increasingly disillusioned with the United States. U.S. support to Pakistan increased Karzai's cynicism about U.S. intentions, alarmed India, and reinforced Pakistan's incentives to allow Taliban sanctuary. The U.S. approach 'became a *de facto* military attrition campaign,' former senior White House official Doug Lute recalls, 'the political and diplomatic efforts never materialized.'[49]

Bureaucratic frictions

Bureaucratic frictions undermined the U.S. government's ability to adapt to a dynamic situation. The Obama administration's strategy called for an annual review. The interpretations of 2010 data were hotly contested. The intelligence community assessed the Taliban had strengthened; the military command countered that the increases in violence were due to ISAF taking the fight into more Taliban controlled areas.[50] The Defence Department insisted the counterinsurgency campaign was on track and simply needed more time to work.[51] After all, the surge forces had only been fully on the ground for a few months. Secretary of Defence Robert Gates noted, 'The sense of progress among those closest to the fight is palpable.'[52]

Holbrooke, who died suddenly of a heart attack during the process, was convinced that COIN had already failed and wanted a major push on reconciliation. Members of the White House staff, sceptical of the surge in the first place, were convinced the military campaign was unlikely to produce decisive victory results.[53] The Departments of Defence and State alike accused the White House staff of being policy advocates rather than honest brokers and of placing the greatest weight on the most negative interpretation of events.[54] Interagency disagreement and an inability to develop and assess strategically relevant metrics reinforced a bias towards maintaining the status quo.

Near the end of 2010, the Obama administration specified five lines of effort: a civil-military campaign to degrade the Taliban and build Afghan

[48]Transparency International, *2010 Corruption Perceptions Index*, https://www.transparency.org/cpi2010/results.
[49]Interview with Douglas E. Lute.
[50]Elizabeth Bumiller, 'Intelligence Reports Offer Dim View of Afghan War', *The New York Times*, 14 December 2010, https://www.nytimes.com/2010/12/15/world/asia/15policy.html.
[51]Vali Nasr, *The Dispensable Nation: American Foreign Policy in Retreat* (New York: Anchor Books 2013), 26, 56–7.
[52]Karen Parrish, 'Gates: Afghanistan Progress Exceeds Expectations', *Armed Forces Press Service* (16 December 2010), http://archive.defense.gov//News/NewsArticle.aspx?ID=62132.
[53]Nasr, *The Dispensable Nation*, 26–7.
[54]Robert M. Gates, *Duty: Memoirs of a Secretary at War* (New York: Alfred A. Knopf 2014), 385; Interviewees H, M, and P, who were involved in the review; my personal recollections as the Department of Defense lead strategist for the policy review.

capacity; strategic partnership; transition to full Afghan sovereignty (security, economic, political); regional diplomacy; and reconciliation.[55] Reflecting the lack of interagency consensus on how to succeed, the lines of effort were co-equal and un-prioritised. President Obama was sticking fast to his 2014 timeline for transition and withdrawal of U.S. forces. The military was tasked in 2009 to hand-over a residual insurgency to the ANSF by 2014, but no standards were set to measure what residual meant or if insurgent strength in 2009 was the benchmark.[56] Elements within the military command, meanwhile, believed that with enough time and pressure on Pakistan, the Taliban could be forced to capitulate (making transition or reconciliation easier). Elements within the State and Defence Departments and White House argued for greater emphasis on reconciliation. Some viewed talks with the Taliban as a way to ease pressures on transition and withdrawal, others believed an effective transition was implausible and a negotiated outcome was the most realistic approach.[57]

In short, advocates for each approach were able to continue their efforts. The result was incoherence. Transition never received the necessary prioritisation and resources to address the problems of legitimacy and sanctuary. Reconciliation lacked political backing in Washington, D.C. and Kabul. The hopes of some that military operations could force the Taliban to sue for peace remained unrealistic.

As the drawdown of U.S. forces began in 2011, the Afghan government was not showing any evidence of reform. Afghanistan remained among the top 3 most corrupt countries in the world.[58] Karzai resisted efforts to address the problem, faulting the United States.[59] Not even the colossal Kabul bank crisis in which nearly $1 billion disappeared could motivate the Obama administration to develop a coherent approach on corruption.[60]

Adding to the problems, insurgent sanctuary in Pakistan remained intact. The efforts to build a strategic partnership with Pakistan that would change the latter's strategic calculus were ineffective. In May 2011, Obama approved the raid into Abbottabad, Pakistan, that killed al Qaeda leader Osama bin Laden. The Pakistan government was outraged by the breach of sovereignty,

---

[55]The White House, 'Overview of the Afghanistan and Pakistan Annual Review' (16 December 2010), https://obamawhitehouse.archives.gov/the-press-office/2010/12/16/overview-afghanistan-and-pakistan-annual-review.

[56]Interviewee M.

[57]As DoD's senior advisor on Afghanistan and Pakistan, I was of the view that a negotiated outcome was the most realistic way of gaining a favorable and durable outcome.

[58]Transparency International, *2011 Corruption Perceptions Index*, https://www.transparency.org/cpi2011/results.

[59]Alissa J. Rubin, 'Karzai Says Foreigners Are Responsible for Corruption', *The New York Times*, 11 December 2011, https://www.nytimes.com/2011/12/12/world/asia/karzai-demands-us-hand-over-afghan-banker.html.

[60]Chayes, *Thieves of State*, 149–54.

ignoring that the world's top terrorist leader was living openly in Pakistan.[61] A final blow came in a border incident in which U.S. aircraft killed twenty-five Pakistan Frontier Corps soldiers who had fired on a nearby ISAF-ANSF patrol.[62] At that point Pakistan cut the ISAF logistics line leading from Karachi to Afghanistan, forcing the coalition to move supplies through Russia and Central Asia instead.

According to officials present for NSC meetings, the Defence and State Departments expressed their concerns about the transition timeline, but not the efficacy of the overall approach.[63] Senior officials discussed risk during Congressional testimonies but did not assess whether the challenges to a successful transition were insurmountable.[64] Members of Congress did not press them on the issue. What was lacking in these discussions was a holistic view of how problems of government legitimacy and the Taliban's internal support and external sanctuary were undermining the prospects of a successful transition. As late as October 2014, the military offered no reason in its semi-annual reports to Congress or testimonies to be overly concerned whether the insurgency would be sufficiently degraded to be handled by the ANSF. 'The ANSF are on track to assume full security responsibility by the end of 2014,' DoD assessed, 'after successfully securing the presidential and provincial council elections and performing well during the fighting season.'[65]

The NSC was unable to address these problems or examine their implications for the transition. Part of the issue was bandwidth. Senior officials complained that getting any time on the National Security Advisor or President's agenda was a tremendous challenge.[66] By 2011, although Deputies meetings were frequent, NSC meetings on Afghanistan were increasingly rare.[67] Major events such as the Arab Spring, Iraq withdrawal, Osama bin Laden raid, Libya intervention, and major domestic challenges all competed for NSC attention. Afghanistan usually reached the president's desk for crisis management.[68] Little time or energy was available for considering highly complex issues, such as whether the U.S. needed to modify the transition timeline or to seek a negotiated outcome instead.

---

[61] Nicholas Schmidle, 'Getting Bin Laden: What happened that night in Abbottabad', *The New Yorker*, 8 August 2011, www.newyorker.com/magazine/2011/08/08/getting-bin-laden.

[62] Salman Masood and Eric Schmitt, 'Tensions Flare Between U.S. and Pakistan After Strike', *The New York Times*, 26 November 2011, www.nytimes.com/2011/11/27/world/asia/pakistan-says-nato-helicopters-kill-dozens-of-soldiers.html.

[63] Interviewees L, M, N, P, Q, W, X.

[64] See the Senate Armed Services Committee, testimonies by Secretary of Defense Leon Panetta and Chairman of the Joint Chiefs of Staff Admiral Michael Mullen (22 September 2011).

[65] Department of Defense, 'Report on Progress Toward Security and Stability in Afghanistan', 1230 Report to Congress (October 2014), 8.

[66] Interviewees L, M, N, P, Q, W, X.

[67] NSC meetings are chaired by the President. Principals meetings are chaired by either the Vice President or the National Security Adviser.

[68] Interviewees L, M, W.

Reconciling reconciliation

Bringing an insurgency to a negotiated outcome has historically been difficult. James D. Fearon found that only 16 percent of the 55 civil wars fought since 1955 ended with negotiated agreements.[69] Paul *et al* found that only 19 of 71 wars against insurgencies since World War Two resulted in 'mixed outcomes,' 27 percent.[70] Such efforts usually fail due to lack of trust: 'combatants are afraid that the other side will use force to grab power and at the same time are tempted to use force to grab power themselves.'[71] An imposing peace-keeping force or third-party enforcer might make the sides abide by an agreement temporarily, but without sufficient trust the power-sharing arrangement is likely to fail.[72]

The Bush administration tended to view the Taliban as a terrorist group and thus ineligible for negotiations.[73] By 2008 an increasing volume of experts and diplomats encouraged a political solution to the conflict.[74] Senior U.S. commanders also began to note that there was no military solution to the conflict.[75] Other experts believed that the surge needed more time to pressure the Taliban and convince them to opt out of the insurgency.[76]

Reconciliation advocates had to contend with the fact that neither the Afghan government nor the Taliban were willing or able to enter peace talks. Powerful constituencies on both sides believed that they could – or must – win outright. For any reconciliation process to become sustainable, these internal groups needed to be brought along to support it. With the conflict in Afghanistan raging for over 30 years by 2010, anxieties and animosities were intense. The Taliban, moreover, knew they could play for time. Obama's announcement of a drawdown timeline signalled to the Taliban that pressure would begin to ease by July 2011.

At least four competing concepts for reconciliation were at work within the U.S. government. The prevailing view, outlined by Secretary of State Hillary Clinton in a speech at the Asia Society in February 2011, was dignified surrender. Taliban who agreed to 'break ties with al-Qaida, give up your

[69]James D. Fearon, 'Iraq's Civil War', *Foreign Affairs* 86/2 (March/April 2007), 2–15.
[70]Paul et al., *Paths to Victory*, 16–21.
[71]Fearon, 'Iraq's Civil War', 2–15.
[72]Barbara Walter, 'The Critical Barrier to Civil War Settlement', *International Organization* 51/3 (Summer 1997), 335–364.
[73]Dobbins and Malkasian, 'Time to Negotiate in Afghanistan', 53–64.
[74]See for instance, Rubin and Rashid, 'From Great Game to Grand Bargain', 30–44; Mohammad Masoom Stanekzai, *Thwarting Afghanistan's insurgency: a pragmatic approach towards peace and reconciliation*, US Institute of Peace special report (September 2008); Adam Roberts, 'Doctrine and Reality in Afghanistan', *Survival* 51/1 (February–March 2009), 29–60.
[75]Radio Free Europe/Radio Liberty, 'Interview: McChrystal Says Solution in Afghanistan Is Developing Governance' (30 August 2009), www.rferl.org/a/Interview_US_Commander_In_Afghanistan_Says_Real_Solution_Is_Developing_Governance/1765881.html.
[76]For example, Sarah Chayes, 'What Vali Nasr gets wrong', *Foreign Policy* (12 March 2013).

arms, and abide by the Afghan constitution [could] rejoin Afghan society; [those] who refus[ed] [would] face the consequences of being tied to al-Qaida as an enemy of the international community.'[77] A second interpretation, which some attributed to Special Representative for Afghanistan and Pakistan Ambassador Richard Holbrooke, was a U.S.-brokered peace deal. A third, which was the position outlined by Holbrooke's successor Marc Grossman, was modest but a more feasible effort designed to broker a meeting between the Afghan government and the Taliban. A fourth position held by some experts in Defence and State (myself included), aimed for a deliberate peace process not unlike that which unfolded in Northern Ireland. U.S. officials often talked past one another using these conflicting definitions.

Making the situation ripe for conflict ending negotiations required the Taliban and Afghan government and their backers to believe that neither side was likely to win outright, that the benefit of future military gains was not worth the cost, and that a path towards a peace process was compelling enough to overcome the status quo.[78] The Obama administration's commitment to withdrawal unwittingly eroded American bargaining leverage with the Afghan government, the Taliban, and Pakistan.

Exploratory talks with the Taliban began in late 2010.[79] By the fall of 2011, the Obama Administration was considering seriously the sequencing of confidence-building measures and wanted to test the waters with Congress. The U.S. Special Representative for Afghanistan and Pakistan explained to members of Congress that the U.S. effort was designed solely to convince the Taliban to meet with the Afghan government. To do so, the Administration would consider trading five Taliban senior leaders in GTMO for one U.S. soldier in Taliban captivity, recognising a Taliban political office, and lifting UN sanctions – all with the aim of brokering a meeting. Members of Congress voiced strong opposition to what came across as a high-risk, low-reward, proposition.[80] Sensitive to such audience costs, the Obama Administration grew even less inclined to make reconciliation the strategic priority.

Tensions emerged with the Afghan government, too. Karzai was sceptical of reconciliation. With enough money and U.S. pressure on Pakistan, Karzai believed, Taliban senior leaders would defect. Karzai was not interested in talks between his government and the insurgency, which he believed gave the latter political legitimacy they did not deserve. He saw the Taliban as

---

[77]U.S. Department of State, 'Remarks by Secretary of State Hillary Clinton Launch of the Asia Society's Series of Richard C. Holbrooke Memorial Addresses' (18 February 2011), https://2009-2017.state.gov/secretary/20092013clinton/rm/2011/02/156815.htm.

[78]For ripeness theory see I. William Zartman, 'The Timing of Peace Initiatives: Hurting Stalemates and Ripe Moments', *The Global Review of Ethnopolitics* 1/1 (September 2001), 8–18.

[79]Steve Coll, *Directorate S: The C.I.A. and America's Secret Wars in Afghanistan and Pakistan* (New York: Penguin Books, 2018), 513–85. (I served as the Secretary of Defense's representative).

[80]Deirdre Walsh and Ted Barrett, 'Congressional leaders initially pushed back on Bergdahl swap', *CNN* (4 June 2014), https://edition.cnn.com/2014/06/03/politics/berdahl-congress-consultation/index.html.

completely dependent on Pakistan. He also grew increasingly concerned that the U.S. was attempting to make a deal with the Taliban as a cover for withdrawal, as they did with North Vietnam in the Paris Peace Accords. In November 2011, Karzai rubbished the idea of a Taliban political office unless it was controlled by the Afghan government. This was a poison pill. Taliban, meanwhile, refused to agree to a travel ban on any detainees transferred from GTMO. This ended discussions on a detainee exchange. The Taliban, arguing that the U.S. was negotiating in bad faith, suspended talks in March 2012.

In early 2013 efforts to open the Taliban office began anew. The Taliban were willing to make another go at it. Karzai asked for a letter from President Obama assuring that the office would not refer to itself as 'the Islamic Emirate of Afghanistan' or look or act as an Embassy. Obama provided it.[81] The U.S., however, failed to ensure all parties agreed on the terms of the office and the effort was a disaster. On 18 June 2013, the office opened, situated in a large enclosed compound nearby embassies. The Qatar-based *al Jazeera* broadcast the opening live on international television; a large banner behind the speakers declared the opening of the 'Political Office of the Islamic Emirate of Afghanistan.' A senior Afghan official considered this a 'breach of the written assurances we received from the U.S. government.'[82]

The poorly coordinated effort to make progress on reconciliation had adverse consequences on the civil-military campaign, transition, the bi-lateral security agreement (BSA), and regional diplomacy – not to mention American credibility. 'The reconciliation effort became too limited,' former Deputy SRAP Singh observed, 'and was not fully connected to the strategy.'[83] Months later, at the *Loya Jirga* designed to discuss the Bi-Lateral Security Agreement (BSA) with the United States for post-2014 troop presence, Karzai took the podium and explained to the delegates that he would not sign the agreement. He noted how assurance letters from Obama could not be trusted.

The Taliban, for their part, were not interested in negotiations to end the conflict. The group's participation in talks reflected a strategic calculation. Their maximalist objective may have been to negotiate U.S. withdrawal: they would cut ties with al Qaeda if the U.S. cut ties with Karzai. At a minimum, they wanted to gain political legitimacy, build relationships, and explore ways to cooperate with the United States without undermining their overall prospects of success.

---

[81]The letter's key assurances were confirmed by a White House official to *The New York Times*. Alissa J. Rubin and Rod Nordlund, 'U.S. Scrambles to Save Taliban Talks After Afghan Backlash', *The New York Times*, 19 June 2013, www.nytimes.com/2013/06/20/world/asia/taliban-kill-4-americans-after-seeking-peace-talks.html.

[82]Karen DeYoung, Tim Craig and Ernesto Londoño, 'Despite Karzai's ire, U.S. confident that talks with Taliban will be held', *The Washington Post*, 19 June 2013, www.washingtonpost.com/world/taliban-says-it-killed-4-us-troops/2013/06/19/512291f0-d8a2-11e2-b418-9dfa095e125d_story.html?utm_term=.43bfc3010b68.

[83]Interview with Vikram Singh.

If the war stalemated, their leverage in negotiations with the Afghan govern-ment would be far higher after international forces left.

Lack of vision and strategy, poor coordination, and sloppy execution reinforced these obstacles and friction points, dooming the reconciliation effort to failure. Success might have been a long-shot, as two former SRAPs have argued.[84] These problems made the odds far steeper than they needed to be. 'The United States and Afghan governments,' Lute observes, 'squan-dered their best opportunity to advance reconciliation.'[85]

Conclusion

The Afghanistan case study holds important lessons on the challenges of successfully concluding irregular conflicts. Although important details differ, the general patterns are similar to those in Vietnam and Iraq.[86] With no doctrine or conceptual apparatus for war termination, the United States repeatedly fails to consider options beyond decisive military victory for achieving a favourable and durable outcome. For irregular wars in which an insurgency has internal support and access to sanctuary, and in which the host nation government is unable to win the battle of legitimacy in insurgent controlled and contested areas, the likelihood of decisive military victory is negligible.[87] The failure to examine other war-termination options during policy and strategy development puts the United States at extreme risk of undertaking large-scale commitments with very limited prospects of success.

The U.S. government then has difficulty modifying a losing or ineffective strategy due to cognitive bias, political and bureaucratic frictions, and challenges with the host nation government. As these problems drag out the conflict, the U.S. government pays penalties in public support. When the United States decides to withdraw, its bargaining leverage is undermined. At this point, the United States is unable to organise incentives sufficiently compelling for the host nation to reform itself enough to win or for the insurgency to negotiate an end to the conflict.

The U.S. government's incoherence on war termination has proven costly. Both administrations tended to focus on the wrong problems – hunting insurgents rather than dealing with the critical factors of insurgent local support and sanctuary, and the host nation government's ability to win the battle of legitimacy in insurgent controlled and contested areas. Failure to address these critical factors doomed Obama's transition and withdraw plan. The so-called crossover point, at which international capacity building

---

[84]Mark Grossman, 'Talking to the Taliban 2010–2011: A Reflection', *Prism* 4/4 (2013), 21–37; Dobbins and Malkasian, 'Time to Negotiate in Afghanistan', 53–64.
[85]Interview with Douglas E. Lute.
[86]Kolenda, *End-Game*.
[87]Kolenda, *End-Game*, 27; Paul et al., *Paths to Victory*, xxi – xxvii, 18.

efforts enabled the Afghan government and security forces to overmatch the Taliban, remained elusive.

Meanwhile, the Obama administration experimented with ways to advance a negotiated outcome – what was termed 'reconciliation.' Reconciliation, however, could not pass the credibility bar with Obama's NSC. The effort had to be sufficiently compelling to overcome the status quo bias of the key actors, their fears of perfidy and loss, and major audience costs. It did none of that and wound up creating antibodies within Afghanistan, regional actors, the Taliban, and even within the U.S. government.

The absence of a conceptual apparatus within the U.S. government for winning wars undermined strategic decision-making. A drawdown timeline, General Petraeus reflected, 'tells the enemy that he just has to hang on for a certain period and then the pressure will be less. In a contest of wills, that matters.'[88] The hopes that the timeline would incentivise Afghan government political reform and undercut the Taliban's strategic narrative were insufficiently scrutinised.

This article has a number of important implications for U.S. national security decision-making. First, war termination outcomes must be imbedded in doctrine and used in policy and strategy-making. Second, the U.S. needs to develop a body of expert knowledge about transition and negotiated outcomes. Having useful case studies and longitudinal analyses can improve decision-making. Third, assessments need to focus on critical strategic factors – in this case, local and external support for the insurgency and the ability of the host government to win the battle of legitimacy. The direction and strength of these factors should play a key role in policy and strategy development. Finally, the U.S. needs to find more effective ways to integrate interagency efforts on the ground. Micromanagement from Washington, DC has proven ineffective time and again. Placing all deployed elements of national power under a single civilian or military authority can reduce the problems of bureaucratic silos and cognitive bias. In short, the U.S. has an opportunity to learn from the Afghanistan experience and to address national security challenges in irregular wars more effectively.

Acknowledgements

With sincere appreciation to Chiara Libiseller and Caroline Bechtel for their research and editorial support.

Disclosure statement

No potential conflict of interest was reported by the authors.

---

[88]Interview with General David H. Petraeus.

Bibliography

Barfield, Thomas, *Afghanistan: A Cultural and Political History* (Princeton: Princeton University Press 2010).

Brahimi, Lakhdar, 'State Building in Crisis and Post-Conflict Countries', 7th Global Forum on Reinventing Government Building Trust in Government, Vienna, Austria, 26–29 June 2007. http://unpan1.un.org/intradoc/groups/public/documents/UN/UNPAN026305.pdf

Chandrasekaran, Rajiv, *Little America: The War within the War for Afghanistan* (New York: Vintage 2012).

Chayes, Sarah, 'What Vali Nasr Gets Wrong', *Foreign Policy*, 12 Mar. 2013. http://foreignpolicy.com/2013/03/12/what-vali-nasr-gets-wrong/

Chayes, Sarah, *Thieves of State: Why Corruption Threatens Global Security* (New York: W.W. Norton 2015).

von Clausewitz, Carl, *On War*, edited and translated by Michael Howard and Peter Paret (Princeton, NJ: Princeton University Press 1984).

Coll, Steve, *Directorate S: The C.I.A. And America's Secret Wars in Afghanistan and Pakistan* (New York: Penguin Books 2018).

Derksen, DeeDee, *The Politics of Disarmament and Rearmament in Afghanistan* (U.S. Institute of Peace May 2015). https://www.usip.org/publications/2015/05/politics-disarmament-and-rearmament-afghanistan

Dobbins, James and Carter Malkasian, 'Time to Negotiate in Afghanistan: How to Talk to the Taliban', *Foreign Affairs*, 94/4 (July/August 2015), 53–64.

Dobbins, James, John G. McGinn, Seth G. Keith Crane, Rollie Jones, Andrew Rathmell Lal, M. Swanger Rachel, and Anga R. Timilsina, *America's Role in Nation-Building: From Germany to Iraq* (Santa Monica, CA: RAND Corporation 2003).

Fänge, Anders, 'The Emergency Loya Jirga', in Martine van Bijlert and Sari Kouvo (eds.), *Snapshots of an Intervention: The Unlearned Lessons of Afghanistan's Decade of Assistance (2001–2011)* (Afghan Analysts Network 2012), 13–170. https://www.afghanistan-analysts.org/wp-content/uploads/downloads/2012/09/Snapshots_of_an_Intervention.pdf

Farrell, Theo and Michael Semple, 'Making Peace with the Taliban', *Survival* 57/6 (December 2015–January 2016), 79–110. doi:10.1080/00396338.2015.1116157

Fearon, James D., 'Iraq's Civil War', *Foreign Affairs* 86/2 (March/April 2007), 2–15.

Felbab-Brown, Vanda, 'Slip-Sliding on a Yellow Brick Road: Stabilization Efforts in Afghanistan', *Stability: International Journal of Security and Development* 1/1 (2012), 4–19. doi:10.5334/sta.af

Gates, Robert M., *Duty: Memoirs of a Secretary at War* (New York: Alfred A. Knopf 2014).

Giustozzi, Antonio, 'The Changing Nature of the Insurgency', *ISAF Strategic Assessment Capability Final Workshop*, Kabul, Afghanistan *10–12 Dec 2014, Ver. 0.1*, Jan. 2015.

Giustozzi, Antonio, *The Army of Afghanistan: A Political History of A Fragile Institution* (London: Hurst 2016).

Goodhand, Jonathan and Aziz Hakimi, *Counterinsurgency, Local Militias, and Statebuilding in Afghanistan* (United States Institute of Peace January 2014), https://www.usip.org/sites/default/files/PW90-Counterinsurgency-Local-Militias-and-Statebuilding-in-Afghanistan.pdf.

Gopal, Anand, *No Good Men among the Living: America, the Taliban, and the War through Afghan Eyes* (New York: Metropolitan Books 2014).

Grossman, Marc, 'Talking to the Taliban 2010-2011: A Reflection', *Prism* 4/4 (2013), 21–37.

Human Rights Watch, 'Violent Response: The U.S. Army in al-Falluja,' 16 June 2003, https://www.hrw.org/report/2003/06/16/violent-response/us-army-al-falluja.

Iklè, Fred C., *Every War Must End* (New York: Columbia University Press 1991).

Katzman, Kenneth, 'Afghanistan: Post-War Governance, Security, and U.S. Policy', Congressional Research Service (28 December 2004), https://www.globalsecurity.org/military/library/report/crs/33748.pdf.

Khalilzad, Zalmay, *The Envoy: From Kabul to the White House, My Journey through a Turbulent World*, (New York: St. Martin's Press, Kindle Edition 2016)

Kolenda, Christopher D., 'End-Game: Why American Interventions Become Quagmires', Doctoral Thesis, King's College London, 2017).

Kolenda, Christopher D., 'Winning Afghanistan at the Community Level', *Joint Force Quarterly*, 56 (1st Quarter 2010), 25–31.

Kolenda, Christopher D., Rachel Reid, and Christopher Rogers, *The Strategic Costs of Civilian Harm: Applying Lessons from Afghanistan to Current and Future Conflicts* (Open Society Foundations June 2016), https://www.opensocietyfoundations.org/sites/default/files/strategic-costs-civilian-harm-20160622.pdf.

Kolhatkar, Sonali and James Ingalls, *Bleeding Afghanistan: Washington, Warlords, and the Propaganda of Silence* (New York: Seven Stories 2006).

Lee, Bradford A., 'Winning the War but Losing the Peace? the United States and the Strategic Issues of War Termination', in Bradford A. Lee and F. Walling Karl (eds.), *Strategic Logic and Political Rationality* (London: Frank Cass 2003), 249–73.

Lee, Bradford A. and Karl F. Walling (eds.), *Strategic Logic and Political Rationality: Essays in Honor of Michael I. Handel* (London: Frank Cass 2003).

Lee, Bradford A. and Karl F. Walling, 'Introduction', in ibid (eds.), *Strategic Logic and Political Rationality* (London, Frank Cass, 2003), 1–27.

Manea, Octavian and John A. Nagl, 'COIN Is Not Dead: An Interview with John Nagl', *Small Wars Journal* (6 February 2012).

McCaffrey, Barry R., *Academic Report: Trip to Afghanistan and Pakistan* (United States Military Academy Department of Social Sciences 3 June 2006), https://www.mccaffreyassociates.com/uploads/pdf/AfghanAAR-052006.pdf.

McChrystal, Stanley A., *COMISAF's Initial Assessment* (Kabul, Afghanistan: Headquarters, International Security Assistance Forces 30 August 2009).

Moten, Matthew (ed.), *Between War and Peace: How America Ends Its Wars* (New York: Free Press 2011).

Nasr, Vali, *The Dispensable Nation: American Foreign Policy in Retreat* (New York: Anchor Books 2013).

NATO, *Progress in Afghanistan, Bucharest Summit* (2–4 April 2008), https://www.nato.int/nato_static/assets/pdf/pdf_publications/progress_afghanistan_2008.pdf.

O'Brien, Paul and Paul Baker, 'Old Questions Needing New Answers: A Fresh Look at Security Needs in Afghanistan', *Paper presented at the Bonn e-conference on Afghanistan: 'Assessing the Progress of Security Sector Reform*, (Bonn, Germany, 4–11 June 2003.

Osman, Borhan and Kate Clark, 'Who Played Havoc with the Qatar Talks? Five Possible Scenarios to Explain the Mess', (Afghan Analysts Network 9 July 2013), https://www.afghanistan-analysts.org/who-played-havoc-with-the-qatar-talks-five-possible-scenarios-to-explain-the-mess/.

Paul, Christopher, Colin P. Clarke, Beth Grill, and Molly Dunigan, *Paths to Victory: Lessons from Modern Insurgencies* (Washington DC: RAND 2013).

Perito, Robert M., *Afghanistan's Police: The Weak Link in Security Sector Reform* (Washington, DC: United States Institute of Peace August 2009).

Rice, Condoleezza, *No Higher Honor: A Memoir of My Years in Washington* (New York: Random House 2011).

Roberts, Adam, 'Doctrine and Reality in Afghanistan', *Survival* 51/1 (February–March 2009), 29–60. doi:10.1080/00396330902749673

Rose, Gideon, *How Wars End: Why We Always Fight the Last Battle* (New York: Simon and Schuster 2010).

Rubin, Barnett R., 'Saving Afghanistan', *Foreign Affairs* 86/1 (January/February 2007), 57–78.

Rubin, Barnett R., *Afghanistan from the Cold War through the War on Terror* (New York: Oxford University Press 2013).

Rubin, Barnett and Ahmed Rashid, 'From Great Game to Grand Bargain: Ending Chaos in Afghanistan and Pakistan', *Foreign Affairs* 87/6 (November/December 2008), 30–44.

Rumsfeld, Donald H., *Known and Unknown: A Memoir* (New York: Sentinel 2011).

Thomas, Ruttig. May 2011. 'The Battle for Afghanistan: Negotiations with the Taliban: History and Prospects for the Future,' National Security Studies Program Policy Paper, *New America Foundation*, 1-30.

Ruttig, Thomas, 'Flash to the Past: Power Play before the 2002 Emergency Loya Jirga', (Afghan Analysts Network 27 April 2012), https://www.afghanistan-analysts.org/flash-to-the-past-power-play-before-the-2002-emergency-loya-jirga/.

Sedra, Mark, 'Police Reform in Afghanistan: An Overview', *Paper presented at the Bonn International Center for Conversion (BICC) e-conference on Afghanistan: 'Assessing the Progress of Security Sector Reform, One Year After the Geneva Conference*, Bonn, Germany, 4–11 June 2003.

Singh, Danny, 'Explaining Varieties of Corruption in the Afghan Justice Sector', *Journal of Intervention and Statebuilding*, 9/2 (May 2015), 231–55. doi:10.1080/17502977.2015.1033093

Special Inspector General for Afghanistan Reconstruction (SIGAR), *Reconstructing the Afghan National Defense and Security Forces: Lessons from the U.S. Experience in Afghanistan* (Washington, DC: U.S. Government Printing Office 2017).

Stanekzai, Mohammad Masoom, 'Thwarting Afghanistan's Insurgency: A Pragmatic Approach Towards Peace and Reconciliation', Special report, Sept. (US Institute of Peace 2008), https://www.usip.org/sites/default/files/sr212.pdf.

Stanley, Elizabeth A. and John P. Sawyer, 'The Equifinality of War Termination Multiple Paths to Ending War', *Journal of Conflict Resolution* 53/5 (October 2009), 651–76. doi:10.1177/0022002709343194

Department of Defense, 'Report on Progress Toward Security and Stability in Afghanistan', 1230 Report to Congress, Oct. (2014), https://dod.defense.gov/Portals/1/Documents/pubs/Oct2014_Report_Final.pdf.

Walter, Barbara, 'The Critical Barrier to Civil War Settlement', *International Organization* 51/3 (Summer 1997), 335–64. doi:10.1162/002081897550384

Wilder, Andrew, 'A House Divided? Analysing the 2005 Afghan Elections (afghan Research and Evaluation Unit 2005)', https://areu.org.af/wp-content/uploads/2015/12/531E-A-House-Divided-IP-print.pdf.

Woodward, Bob, *State of Denial: Bush at War, Part III* (New York: Simon and Schuster 2006).

William, Zartman, I., 'The Timing of Peace Initiatives: Hurting Stalemates and Ripe Moments', *The Global Review of Ethnopolitics* 1/1 (September 2001), 8–18.

Interviews Cited

Lute, Douglas E. 20 September 2016. Former Deputy National Security Advisor and Assistant to the President for Iraq and Afghanistan (2007 – 2013); former U.S. Ambassador to NATO (2013-2017).

Petraeus, David H., General. 8 December 2016. Former Commander: Multi-National Forces-Iraq (2007-8); U.S. Central Command (2008-10); International Security Assistance Forces – Afghanistan (2010-2011); former Director, Central Intelligence Agency (2011-3).

Rubin, Barnett R. 12 September 2016. Afghanistan expert. Former senior advisor to U. S. Special Representative for Afghanistan and Pakistan (2009-2014).

Vikram, Singh. 7 September 2016. Former Deputy to U.S. Special Representative for Afghanistan and Pakistan (2009-2011).

Interviews Cited on Condition of Anonymity

Interviewee H: Former State department senior advisor.
Interviewee J: Former State Department senior advisor.
Interviewee L: A former senior White House official in Obama Administration.
Interviewee M: A former senior Obama administration official.
Interviewee N: A former senior ISAF official.
Interviewee P: A former senior ISAF official.
Interviewee Q: A former senior ISAF official.
Interviewee S: A former commander in the Afghan security forces.
Interviewee W: Former senior Pentagon official in the Obama administration.
Interviewee X: Former senior Pentagon official in the Obama administration.
Interviewee Y: Former Taliban senior official.

Index

Note: Endnotes are indicated by the page number followed by "n" and the endnote number e.g., 57n50 refers to endnote 50 on page 57.